AF771133

MECHANICAL ENGINEERING THEORY AND APPLICATIONS

MODEL PREDICTIVE CONTROL

THEORY, PRACTICES AND FUTURE CHALLENGES

Mechanical Engineering Theory and Applications

Additional books in this series can be found on Nova's website under the Series tab.

Additional e-books in this series can be found on Nova's website under the e-books tab.

MECHANICAL ENGINEERING THEORY AND APPLICATIONS

MODEL PREDICTIVE CONTROL

THEORY, PRACTICES AND FUTURE CHALLENGES

CORRINE WADE
EDITOR

nova publishers
New York

Copyright © 2015 by Nova Science Publishers, Inc.

All rights reserved. No part of this book may be reproduced, stored in a retrieval system or transmitted in any form or by any means: electronic, electrostatic, magnetic, tape, mechanical photocopying, recording or otherwise without the written permission of the Publisher.

We have partnered with Copyright Clearance Center to make it easy for you to obtain permissions to reuse content from this publication. Simply navigate to this publication's page on Nova's website and locate the "Get Permission" button below the title description. This button is linked directly to the title's permission page on copyright.com. Alternatively, you can visit copyright.com and search by title, ISBN, or ISSN.

For further questions about using the service on copyright.com, please contact:
Copyright Clearance Center
Phone: +1-(978) 750-8400 Fax: +1-(978) 750-4470 E-mail: info@copyright.com.

NOTICE TO THE READER

The Publisher has taken reasonable care in the preparation of this book, but makes no expressed or implied warranty of any kind and assumes no responsibility for any errors or omissions. No liability is assumed for incidental or consequential damages in connection with or arising out of information contained in this book. The Publisher shall not be liable for any special, consequential, or exemplary damages resulting, in whole or in part, from the readers' use of, or reliance upon, this material. Any parts of this book based on government reports are so indicated and copyright is claimed for those parts to the extent applicable to compilations of such works.

Independent verification should be sought for any data, advice or recommendations contained in this book. In addition, no responsibility is assumed by the publisher for any injury and/or damage to persons or property arising from any methods, products, instructions, ideas or otherwise contained in this publication.

This publication is designed to provide accurate and authoritative information with regard to the subject matter covered herein. It is sold with the clear understanding that the Publisher is not engaged in rendering legal or any other professional services. If legal or any other expert assistance is required, the services of a competent person should be sought. FROM A DECLARATION OF PARTICIPANTS JOINTLY ADOPTED BY A COMMITTEE OF THE AMERICAN BAR ASSOCIATION AND A COMMITTEE OF PUBLISHERS.

Additional color graphics may be available in the e-book version of this book.

Library of Congress Cataloging-in-Publication Data

Model predictive control : theory, practices and future challenges / [edited by] Corrine Wade.
pages cm. -- (Mechanical engineering theory and applications)
Includes bibliographical references and index.
ISBN 978-1-63463-859-3 (hardcover)
1. Predictive control. I. Wade, Corrine, editor.
TJ217.6.M63 2015
629.8--dc23
2014048429

Published by Nova Science Publishers, Inc. † New York

CONTENTS

PREFACE

Although industrial processes are inherently nonlinear, many contributions for controller design for those plants are based on the assumption of a linear model of the system. However, in some cases it is difficult to represent a given process using a linear model. Model Predictive Control (MPC) is an optimal control approach which can effectively deal with constraints and multivariable processes in industries. Because of its advantages, MPC has been widely applied in automotive and process control communities. This book discusses the theory, practices and future challenges of model predictive control.

Among the several predictive control approaches proposed in the literature (hysteresis-based, trajectory-based, deadbeat, etc.), Model Predictive Control (MPC) surely represents the most promising one due to its inherent flexibility and versatility. In fact, MPC cost function generally consists of a sum of several terms, whose weights can be tuned in accordance with different criteria. As a result, an accurate choice of MPC cost function enables an appropriate optimization of several non-linear systems (electrical, mechanical, chemical, etc.), especially those characterized by several inputs and/or outputs. One of the most important feature of MPC consists of more easily taking into account both input and output constraints compared to other predictive control approaches, because they can be accounted by the MPC cost function. However, this solution does not guarantee that they are always satisfied, may impairing system performance optimization at the same time.

A viable solution consists of appropriately managing input and output constraints in order to guarantee system optimization. In particular, system operating boundaries should be determined at first, within which a number of MPC objective functions should be minimized in a descending order of priority. Although such an approach is generally quite complex compared to conventional MPC, it leads to a better system exploitation. In addition, higher computational efforts can be easily handled by means of fast processing units, like Field Programmable Gate Arrays (FPGA); in fact, they allow very fast execution times, even for advanced MPC-based control systems, making them particularly suitable in replacing traditional control techniques in industrial applications.

Chapter 1 addresses the problem of accurate management of input and output constraints for MPC. Thus, firstly referring to a generic system, problem statement and formulation are firstly introduced and briefly discussed. Subsequently, reference is made to a case study, namely a Surface-Mounted Permanent Magnet Synchronous Machine (SPM). In particular, its mathematical modelling is briefly introduced at first, as well as its operating constraints (voltage saturation, current limitations, etc.). Subsequently, the design of an MPC algorithm is reported, which is based on accurate management of SPM operating constraints in order to guarantee optimal SPM performances, over both steady-state and dynamic operation. Both simulation and experimental results are also enclosed in order to highlight the effectiveness of this MPC approach; in particular, simulations are carried out by means of Matlab-Simulink, whereas experiments refer to the employment of an appropriate FPGA-based control board.

Chapter 2 deals with a hybrid actuator composed by a piezo and a hydraulic part and with a Robust Model Predictive Control (RMPC) structure combined with a feedforward control in camless engine motor applications. A combination between a feedforward control based on an inversion of the system and an MPC structure is considered. To perform a feedforward regulator an identification of the start condition of the piezo actuator is needed. This start condition of the piezo actuator is due to some structural constructive aspects which generate an offset into the piezo position. The feedforward regulator ends up being an affine function to compensate for this offset. A procedure for its identification is proposed. The idea behind the conception of the proposed new actuator is to use the advantages of both the high precision of the piezo and the force of the hydraulic part. In fact, piezoelectric actuators (PEAs) are commonly used for precision positionings, despite the fact that PEAs present nonlinearities, such as hysteresis, saturations, and creep. In the control problem such nonlinearities must be taken into account. In this paper the Preisach dynamic model with the abovementioned nonlinearities is considered together with a feedforward control combined with a RMPC. Simulations of the implementation of the MPC structure together with the feedforward regulator and the abovementioned start condition of the piezo actuator with real data are shown.

Mobile robotics is a notable case of such evolution. The robotics community has developed sophisticated analysis and control techniques to meet increasing requirements for the control of motions of mechanical systems. These increasing requirements are motivated by higher performance specifications, notably an increasing number of degrees-of-freedom. Chapter 3 proposes a controller for the motion of the Robotino. The proposed controller takes under consideration a non-interacting control strategy realized using a geometric approach. Horizontal, vertical and angular motions are considered and once the decoupling between these motions is obtained, a Model Predictive Control (MPC) strategy is used in combination with a Feedforward controller. The approach used to obtain a decoupling consists of a geometric approach. In the past three decades, research on the geometric approach to dynamic systems theory and control has allowed this approach to become a powerful and a thorough tool for the analysis and synthesis of dynamic systems. Simulation results using real data of the Robotino are shown.

There are very few controller design techniques that can be proven to stabilize processes in the presence of nonlinearities and constraints. Model predictive control (MPC) is one of these techniques. For this reason, there has been much interest in nonlinear model-based control within the process engineering community. A critical step in the application of these methods is the development of a suitable model for the process dynamics. In this sense, block-oriented models have proved to be useful as simple nonlinear models for a vast number of applications. They are described as a cascade of linear dynamic and nonlinear static blocks. They have emerged as an appealing proposal due to their simplicity and the property of being valid over a larger operating region than a linear time invariant (LTI) model. A typical block-oriented model found in the literature is the Hammerstein model. In Chapter 4 a nonlinear memoryless block is followed by a linear dynamics. A broad type of dynamic processes can be described by such representations consisting of these two simple elements usually referred to as subsystems. This chapter deals with robust control for uncertain Hammerstein models. The starting point for the controller design is a Hammerstein model which describes the systems dynamics in the presence of uncertainty. This model is employed to design a model based predictive controller. The mathematical problem involved in the development of the algorithm is stated in the context of Linear Matrix Inequalities (LMI) theory. The straightforward use of Hammerstein models for designing the Model Predictive controller would lead to a nonlinear optimization problem due to the static nonlinearity. From the point of view of the implementation, this could result in high computational complexity and be a very time-consuming process. This can be avoided by exploiting the structure of the Hammerstein model, which is a novel approach. This strategy developed in this chapter takes advantage of the static nature of the nonlinearity which allows being transformed into polytopic representation and, therefore, to solve the control problem by focusing only in the linear dynamics. This formulation results in a simplified design procedure, because the original nonlinear Model Predictive Control problem turns into a linear one. At the end of the chapter, different simulation examples are presented to illustrate the controller design procedure.

In model predictive control (MPC) the basic notions are a trajectory and a set of feasible trajectories. By using these notions the author's describe the MPC procedure in Hilbert space. It is shown that a feasible set of trajectories can be associated with a bounded positive self-adjoint linear operator in Hilbert space. This operator gives a complete description of CS (at a given time or a state). In the same way a target of control is described. Chapter 5 shows that the principal stage of MPC - reconstruction of a new trajectory - can be formalized as the Newton transformation of a given trajectory. The author's show how these constructions in Hilbert space can be transformed into a mathematical (identified) model. For this an operator defined by relations between two bases (in Hilbert and finite coordinate spaces) is introduced and called a realization operator. This approach is generalized on a collection of trajectories (in Hilbert and coordinate spaces respectively) and naturally leads to frame theory. A predictive frame that gives a complete description of CS (at a given time or a state) is introduced in Hilbert space. By using the realization operator the author's transform the predictive frame into a table of numbers and we call it data predictive matrix. The data predictive matrix is a basis of a new approach to predictive control which the author's call data predictive control (DPC). The author's analyze the stages of DPC and show how by using different control strategies the construction of data predictive matrix can be adopted to goals of control. Finally, the author's consider a construction of the feasible trajectory which is the closest to a target (but unfeasible) trajectory (this procedure is similar to the consistent reconstruction in frame theory).

Consider a complex large-scale control system which is composed of many spatially distributed subsystems. Each subsystem interacts with some other subsystems by their states and/or inputs, e.g. large-scale chemical process, smart grid, distributed generation systems. The control objective is to achieve a specific global performance of the entire system or a common goal of all subsystems.

To control this class of system, the Distributed Model Predictive Control (DMPC), which controls each subsystem by a separate local Model Predictive Control (MPC), has become more and more popular since it not only inherits MPC's ability to explicitly accommodate constraints but also possesses the advantages of the distributed framework of good flexibility and good error tolerance. On the other hand, with the development of communication network technologies in process industries, which allow a distributed controller to access and send information throughout the system, also helps to promote distributed control solutions. However, as point in many articles, the performance of a DMPC is, in most cases, not as good as that of a centralized MPC.

The flexibility (or error tolerance) and global performance is two important characteristics of a DMPC. To improve the optimization performance, the existing methods usually increase the coordination degree (the range of cost that each subsystem-based MPC minimized). With the increasing of the coordination degree, the performance of entire system becomes better and better. However, with the increasing of the coordination degree, the network connectivity become more and more complicity, and consequently the error tolerance and high flexibility become weaker and weaker. It is not expected. Can the author's find a method which could improve the global performance or coordination degree without any increasing of network connectivity?

In Chapter 6, a novel coordination strategy, where each subsystem-based model predictive control (MPC) added a quadratic function of the affection of the current subsystem's input to its down-stream neighbors into its optimization index, is proposed for improving the optimization performance of entire system. This method is able to increase the coordination degree without any increasing of network connections comparing to the methods which do not use this coordination strategy. The consistency constraints, which limit the error between the state predicted at the previous time instant, referred to as the presumed state, and the state predicted at the current time instant within a prescribed bound, are designed and included in the optimization problem of each subsystem-based MPC. These constraints guarantee the recursive feasibility of each subsystem-based MPC. In the meantime, a stabilization constraint and the dual mode predictive control strategy are adopted to result in a stabilizing DMPC.

A set of feasible trajectories for a control system is analyzed. Chapter 7 considers real trajectories (realized by a given control system or calculated by a computer) and possible trajectories (subjected to the equations and constraints of a given mathematical model). Therefore the author's have in model predictive control (MPC) the space of real trajectories and the space of possible trajectories. It is shown that the basic notion for the space of possible trajectories is an equivalence relation and associated with this relation a set of feasible trajectories (a ball in a Banach space). This ball with some additional requirements of symmetry determines a positive inner product; thus the space of possible trajectories becomes a Hilbert space (an Euclidean space for finite dimensional space). It is shown that the basic notion for the space of real trajectories is an order relation and associated with this relation a positive cone. Under some additional requirements of symmetry the cone becomes an elliptical cone that determines a pseudo-Euclidean metric in the space of real trajectories. In the last part of the paper the space of trajectories is considered as a direct sum of two spaces introduced above. A control possibilities set is introduced in this space. It is shown that the projections of this set onto the above spaces define the structure of control in MPC. This structure is described by the composition of two multivalued mappings: a given trajectory to a set of possible trajectories and a set of possible trajectories to a new (optimal) trajectory called pilot transformation. The author's describe pilot transformation by using the basic constructions of frame theory.

be described by such representations consisting of these two simple elements usually referred to as subsystems. This chapter deals with robust control for uncertain Hammerstein models. The starting point for the controller design is a Hammerstein model which describes the systems dynamics in the presence of uncertainty. This model is employed to design a model based predictive controller. The mathematical problem involved in the development of the algorithm is stated in the context of Linear Matrix Inequalities (LMI) theory. The straightforward use of Hammerstein models for designing the Model Predictive controller would lead to a nonlinear optimization problem due to the static nonlinearity. From the point of view of the implementation, this could result in high computational complexity and be a very time-consuming process. This can be avoided by exploiting the structure of the Hammerstein model, which is a novel approach. This strategy developed in this chapter takes advantage of the static nature of the nonlinearity which allows being transformed into polytopic representation and, therefore, to solve the control problem by focusing only in the linear dynamics. This formulation results in a simplified design procedure, because the original nonlinear Model Predictive Control problem turns into a linear one. At the end of the chapter, different simulation examples are presented to illustrate the controller design procedure.

In model predictive control (MPC) the basic notions are a trajectory and a set of feasible trajectories. By using these notions the author's describe the MPC procedure in Hilbert space. It is shown that a feasible set of trajectories can be associated with a bounded positive self-adjoint linear operator in Hilbert space. This operator gives a complete description of CS (at a given time or a state). In the same way a target of control is described. Chapter 5 shows that the principal stage of MPC - reconstruction of a new trajectory - can be formalized as the Newton transformation of a given trajectory. The author's show how these constructions in Hilbert space can be transformed into a mathematical (identified) model. For this an operator defined by relations between two bases (in Hilbert and finite coordinate spaces) is introduced and called a realization operator. This approach is generalized on a collection of trajectories (in Hilbert and coordinate spaces respectively) and naturally leads to frame theory. A predictive frame that gives a complete description of CS (at a given time or a state) is introduced in Hilbert space. By using the realization operator the author's transform the predictive frame into a table of numbers and we call it data predictive matrix. The data predictive matrix is a basis of a new approach to predictive control which the author's call data predictive control (DPC). The author's analyze the stages of DPC and show how by using different control strategies the construction of data predictive matrix can be adopted to goals of control. Finally, the author's consider a construction of the feasible trajectory which is the closest to a target (but unfeasible) trajectory (this procedure is similar to the consistent reconstruction in frame theory).

Consider a complex large-scale control system which is composed of many spatially distributed subsystems. Each subsystem interacts with some other subsystems by their states and/or inputs, e.g. large-scale chemical process, smart grid, distributed generation systems. The control objective is to achieve a specific global performance of the entire system or a common goal of all subsystems.

To control this class of system, the Distributed Model Predictive Control (DMPC), which controls each subsystem by a separate local Model Predictive Control (MPC), has become more and more popular since it not only inherits MPC's ability to explicitly accommodate constraints but also possesses the advantages of the distributed framework of good flexibility and good error tolerance. On the other hand, with the development of communication network technologies in process industries, which allow a distributed controller to access and send information throughout the system, also helps to promote distributed control solutions. However, as point in many articles, the performance of a DMPC is, in most cases, not as good as that of a centralized MPC.

The flexibility (or error tolerance) and global performance is two important characteristics of a DMPC. To improve the optimization performance, the existing methods usually increase the coordination degree (the range of cost that each subsystem-based MPC minimized). With the increasing of the coordination degree, the performance of entire system becomes better and better. However, with the increasing of the coordination degree, the network connectivity become more and more complicity, and consequently the error tolerance and high flexibility become weaker and weaker. It is not expected. Can the author's find a method which could improve the global performance or coordination degree without any increasing of network connectivity?

In Chapter 6, a novel coordination strategy, where each subsystem-based model predictive control (MPC) added a quadratic function of the affection of the current subsystem's input to its down-stream neighbors into its optimization index, is proposed for improving the optimization performance of entire system. This method is able to increase the coordination degree without any increasing of network connections comparing to the methods which do not use this coordination strategy. The consistency constraints, which limit the error between the state predicted at the previous time instant, referred to as the presumed state, and the state predicted at the current time instant within a prescribed bound, are designed and included in the optimization problem of each subsystem-based MPC. These constraints guarantee the recursive feasibility of each subsystem-based MPC. In the meantime, a stabilization constraint and the dual mode predictive control strategy are adopted to result in a stabilizing DMPC.

A set of feasible trajectories for a control system is analyzed. Chapter 7 considers real trajectories (realized by a given control system or calculated by a computer) and possible trajectories (subjected to the equations and constraints of a given mathematical model). Therefore the author's have in model predictive control (MPC) the space of real trajectories and the space of possible trajectories. It is shown that the basic notion for the space of possible trajectories is an equivalence relation and associated with this relation a set of feasible trajectories (a ball in a Banach space). This ball with some additional requirements of symmetry determines a positive inner product; thus the space of possible trajectories becomes a Hilbert space (an Euclidean space for finite dimensional space). It is shown that the basic notion for the space of real trajectories is an order relation and associated with this relation a positive cone. Under some additional requirements of symmetry the cone becomes an elliptical cone that determines a pseudo-Euclidean metric in the space of real trajectories. In the last part of the paper the space of trajectories is considered as a direct sum of two spaces introduced above. A control possibilities set is introduced in this space. It is shown that the projections of this set onto the above spaces define the structure of control in MPC. This structure is described by the composition of two multivalued mappings: a given trajectory to a set of possible trajectories and a set of possible trajectories to a new (optimal) trajectory called pilot transformation. The author's describe pilot transformation by using the basic constructions of frame theory.

In: Model Predictive Control
Editor: Corrine Wade, pp. 1-39
ISBN: 978-1-63463-859-3
© 2015 Nova Science Publishers, Inc.

Chapter 1

MODEL PREDICTIVE CONTROL WITH INPUT AND OUTPUT CONSTRAINTS

Alessandro Serpi, Gianluca Gatto, Alfonso Damiano and Ignazio Marongiu
Department of Electrical and Electronic Engineering, University of Cagliari, Piazza d'Armi, Cagliari, Italy

Abstract

Among the several predictive control approaches proposed in the literature (hysteresis-based, trajectory-based, deadbeat, etc.), Model Predictive Control (MPC) surely represents the most promising one due to its inherent flexibility and versatility. In fact, MPC cost function generally consists of a sum of several terms, whose weights can be tuned in accordance with different criteria. As a result, an accurate choice of MPC cost function enables an appropriate optimization of several non-linear systems (electrical, mechanical, chemical, etc.), especially those characterized by several inputs and/or outputs. One of the most important feature of MPC consists of more easily taking into account both input and output constraints compared to other predictive control approaches, because they can be accounted by the MPC cost function. However, this solution does not guarantee that they are always satisfied, may impairing system performance optimization at the same time.

A viable solution consists of appropriately managing input and output constraints in order to guarantee system optimization. In particular, system operating boundaries should be determined at first, within which a number of MPC objective functions should be minimized in a descending order of priority. Although such an approach is generally quite complex compared to conventional MPC, it leads to a better system exploitation. In addition, higher computational efforts can be easily handled by means of fast processing units, like Field Programmable Gate Arrays (FPGA); in fact, they allow very fast execution times, even for advanced MPC-based control systems, making them particularly suitable in replacing traditional control techniques in industrial applications.

This chapter addresses the problem of accurate management of input and output constraints for MPC. Thus, firstly referring to a generic system, problem statement and formulation are firstly introduced and briefly discussed. Subsequently, reference is made to a case study, namely a Surface-Mounted Permanent Magnet Synchronous Machine (SPM). In particular, its mathematical modelling is briefly introduced at first, as well as its operating constraints (voltage saturation, current limitations, etc.). Subsequently, the design of an MPC algorithm is reported, which is based on accurate management of SPM operating constraints in order to guarantee optimal SPM performances, over both steady-state and dynamic operation. Both simulation and experimental results are also enclosed in order to highlight the effectiveness of this MPC approach; in particular, simulations are carried out by means of Matlab-Simulink, whereas experiments refer to the employment of an appropriate FPGA-based control board.

Introduction

Considering a generic system, its continuous-time mathematical model can be expressed in terms of state variables as

$$\begin{aligned} \dot{x} &= f(x,u,t) \quad , \quad x(t_0) = x_0 \\ y &= g(x,u,t) \end{aligned} \tag{1}$$

where x and u denote state and input vectors respectively, y being the output vector. In particular, for electrical systems, x generally consists of inductor currents, magnetic flux linkages and/or capacitor voltages, whereas u accounts for voltage and current supplies mostly. Regarding the output vector y, it generally depends on both x and u and consists of those variables whose time-evolution should be imposed appropriately in order to optimize system performances. In this context, apart from (1), a number of input and output constraints have to also be taken into account, which can generally be expressed as:

$$\mathcal{C}_i(x,u,t) \leq 0 \quad , \quad i = 1..n. \tag{2}$$

In particular, input constraints usually account for bounded values of input variables, such as maximum supply voltage and/or current. As a result, input constraints generally affect both dynamic and steady-state system performances, e.g. by increasing response times and preventing the achievement of some operating conditions at the same time. Whereas state and output constraints affect steady-state operations mostly because they account for hazardous conditions that cannot be held continuously. For example, considering an electrical machine, such constraints consist of high currents flowing into electrical machine windings or excessive speed values. The firsts may lead to overheating, which, in turn, may deteriorate winding isolation. Whereas the latter may introduce mechanical problems on shaft and bearings, thus leading to unsuitable vibrations and noise. However, since such operating conditions can be sustained for short periods of time, state and output constraints do not generally affect dynamic system operation significantly.

Thus, based on(1) and (2), or on their corresponding sampled-data versions, several control systems can be designed and implemented in order to achieve system performance optimization, depending on inherent features of the system to be controlled, as well as on specific application needs and requirements. In general, a reference output evolution is imposed (y^*), whose tracking has to be accomplished in accordance with different optimization criteria (fastest response, minimum losses, etc.). Consequently, different optimal control laws (u^*) can be achieved, also in accordance with the kind of control system employed (hysteresis, PI-based, predictive control, etc.). In this context, it is worth noting that hysteresis control systems provide good dynamic responses, being robust to parameter variations and uncertainties too.

Unfortunately, steady-state evolutions are generally characterized by strong ripple, which prevents an optimal tracking of y^*. Differently, PI-based control systems provide good steady-state performances and a fair insensitivity to system parameter variations and uncertainties, but they do not generally guarantee high dynamic performances. Unlike the above mentioned control techniques, predictive control can assure good performances over both dynamic and steady-state operation, but an accurate knowledge of system model and parameters is required. In fact, since predictive control algorithms are generally designed based on sampled-data models of the controlled system, they are badly affected by parameter variations and uncertainties, which may lead to un-optimized system performances. However, this drawback can be overcome by employing appropriate on-line parameter identification and/or adaptive control approaches.

Among the several kinds of predictive control proposed in the literature (hysteresis-based, trajectory-based, deadbeat, etc.), Model Predictive Control (MPC) surely represents the most promising one due to its inherent flexibility and versatility. In fact, referring to Fig. 1, conventional MPC consists in synthesizing the most suitable control law u^* in order to minimize an appropriate cost function, which can be expressed as

$$\Phi(x,u,t,y^*) = \sum_j \lambda_j \cdot \varphi_j(x,u,t,y^*) \quad , \qquad j = 1..m. \tag{3}$$

Particularly, MPC cost function generally consists of a sum of several objective functions, each of which (φ_j) should be minimized in accordance with its corresponding weight (λ_j). As a result, an accurate choice of λ_j enables an appropriate optimization of several non-linear systems, especially those characterized by several inputs and/or outputs. In this context, it is worth noting that different choices of λ_j can lead to different system optimization, making MPC very flexible. In addition, MPC more easily takes into account both input and output constraints compared to other predictive control approaches. In fact, they can be generally included into MPC cost function, leading to

$$\tilde{\Phi}(x,u,t,y^*) = \sum_j \lambda_j \cdot \varphi_j(x,u,t,y^*) + \sum_i \psi_i \cdot \mathcal{C}_i(x,u,t) \quad , \qquad \begin{cases} j = 1..m \\ i = 1..n \end{cases} \tag{4}$$

where each ψ_i denotes the weight of $\mathcal{C}_i$. As a result, the minimization of $\tilde{\Phi}$ instead of Φ should guarantee the achievement of an appropriate control law u^* that is able to track y^*optimally, by satisfying all input and output constraints at the same time.

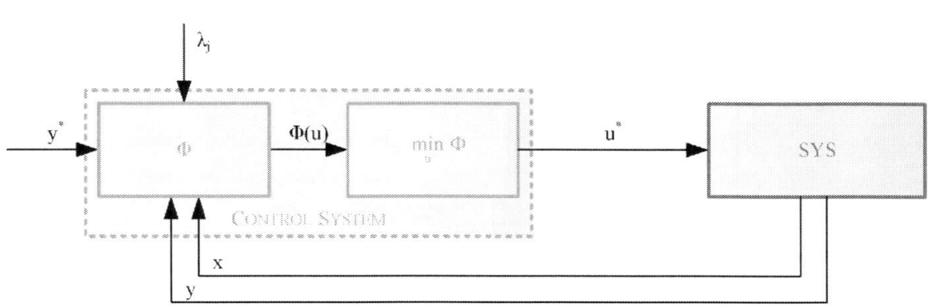

Figure 1. Schematic representation of a conventional Model Predictive Control system.

Although such an MPC approach is quite simple, effective and easy to be implemented, it also reveals some critical issues, which may prevent system performance optimization. In particular, the minimization of Φ defined by (3) always requires a trade-off among φ_j minimizations, each of which can be more or less prioritized by employing high or low λ_j value respectively. Consequently, the employment of $\tilde{\Phi}$ in place of Φ may lead to sub-optimal solutions; in fact, high ψ_i values makes $\tilde{\Phi}$ very sensitive to input and output constraints mostly, thus system performance optimization, which represents the main goal of MPC, may be unsuitably impaired. On the other hand, if ψ_i are chosen relatively low compared to λ_j, $\tilde{\Phi}$ almost equals Φ, thus input and output constraints become quite irrelevant. However, regardless of the choice of ψ_i, the minimization of $\tilde{\Phi}$ as a whole may not comply with all input and output constraints, some of which may be thus not satisfied over both dynamic and steady-state operations.

On the basis of the previous considerations, another MPC approach should be followed in order to assure an appropriate system performance optimization and full compliance with all input and output constraints, over both dynamic and steady-state operation. It consists of carrying out an accurate input and output constraint management at steady-state operation at first, leading to define the following steady-state subset X_0:

$$X_0: \ \forall x \in X_0 \ \rightarrow \ \mathcal{C}_i(x,u,t) \leq 0 \ , \qquad i = 1..n. \tag{5}$$

In particular, since only x within X_0satisfy all input and output constraints, the optimal steady-state solution (x^*) must lie within X_0 for any given y^*. As a result, system performance optimization could be achieved by minimizing Φ within X_0, since there is no need of introducing $\tilde{\Phi}$ further. In addition, assuming that φ_j can be ordered by decreasing priority, several steady-state subset can be introduced as

$$X_j \subseteq X_{j-1}: \ \forall x \in X_j \ \rightarrow \ \varphi_j(x,u,t) = \min_{x \in X_{j-1}} \{\varphi_j(x,u,t)\}, \ j = 1..m. \tag{6}$$

Thus, referring to Fig. 2, it can be stated that the optimal solution x^* can be found without resorting to conventional MPC cost functions, like that expressed by (3). Consequently, the introduction of weights, whose tuning is generally hard to be accomplished and which may lead to unsuitable operating conditions, is not required further. A similar approach can be followed over dynamic operation; in fact, once x^* is determined, its tracking can be accomplished referring to appropriate dynamic subsets ($\tilde{X}$). These last can be defined in accordance with (5) and (6), but based on dynamic operating constraints and objective functions, which may differ from the corresponding steady-state ones.

The design and implementation of this novel MPC approach is shown in the following by referring to a case study, i.e. a Surface-Mounted Permanent Magnet Synchronous Machine (SPM) fed by a three-phase inverter. In particular, SPM mathematical models are briefly introduced at first, as well as its dynamic and steady-state operating constraints. Subsequently, an MPC algorithm is designed based on the above-mentioned approach, whose effectiveness is appropriately highlighted by both simulation and experimental results.

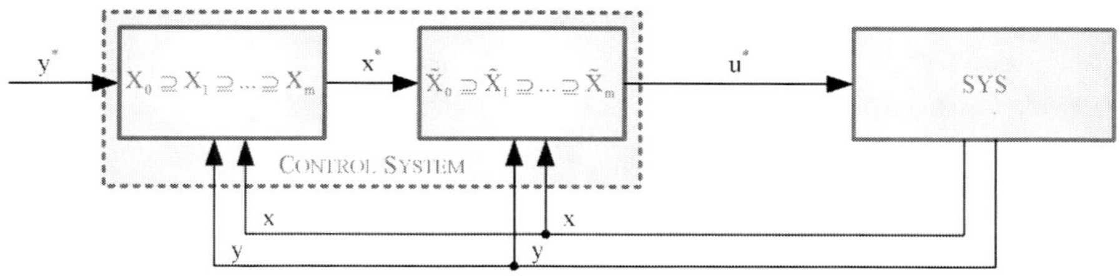

Figure 2. Schematic representation of a novel Model Predictive Control approach.

Surface-Mounted Permanent Magnet Synchronous Machine

Permanent Magnet Synchronous Machines (PMSMs) are nowadays widely employed in a variety of applications, such as electric vehicles, electric ship propulsion, robotics, aerospace and power generation. This is due to their high torque and power density, high efficiency and reliability, which are obtained by employing appropriate permanent magnet materials and machine topologies [1]. Among PMSMs, Surface-Mounted Permanent Magnet Synchronous Machines (SPMs) are very widespread because they are relatively simple to be manufactured and easy to be controlled in comparison with other kinds of PMSMs. However, full exploitation of SPM dynamic and steady-state performances also needs advanced control systems, which should manage SPM operating constraints appropriately, especially at high-speed.

In this context, several predictive control algorithms have been proposed in the literature for power electronic converters and electrical drives [2]-[23], some of which specifically designed for PMSM [12]-[23]. All these aims to enhance dynamic and steady-state performances compared to other control systems. It is worth noting that although the application of predictive control technique for such electrical systems was already suggested a long time ago [2]-[4], its high computational effort prevented it from being widely employed in the past decades. However, it has recently gained more and more interest due to the increasing development of fast processing units, like Field Programmable Gate Arrays (FPGA) [24]-[28]. In particular, FPGAs match the predictive control computational demands well, especially in terms of execution time, even allowing real-time implementation of advanced predictive control algorithms. In addition, FPGAs also enable the implementation of appropriate on-line parameter identification procedures in order to overcome predictive control issues due to parameter variations and uncertainties [29]. However, in spite of this, predictive control algorithms proposed in the literature are generally designed on simplified sampled-data models, as well as on approximated input and output constraints in order to ease algorithm implementation. This prevents the full exploitation of SPM performances, especially above its rated speed. In fact, in this last case, appropriate flux-weakening control strategies have to be employed, which should entail an appropriate management of input and output constraints, over both dynamic and steady-state operation [30]-[46].

On the basis of all the previous considerations, an MPC algorithm for an SPM is presented in the following. In particular, an accurate SPM sampled-data model is introduced at first, which allows the minimization of modelling errors due to the discretization procedure, resulting in enhanced MPC performances [47]. In addition, the MPC algorithm is designed in order to appropriately account for both input and output constraints (current limitation, voltage saturation, etc.), leading to fully exploit SPM performances, over both dynamic and steady-state operation[48]-[50]. The effectiveness of this novel MPC algorithm is validated by means of both simulations and experiments; in particular, simulations are carried out in the Matlab Simulink environment, whereas experiments are performed by means of an FPGA-based control board. These last also regard the comparison between the novel MPC algorithm and a conventional PI-based control system, i.e. the well-known voltage follower based on a PI voltage compensator (VF-PI)[43]-[46], in order to highlight the enhanced SPM performances achievable by the former.

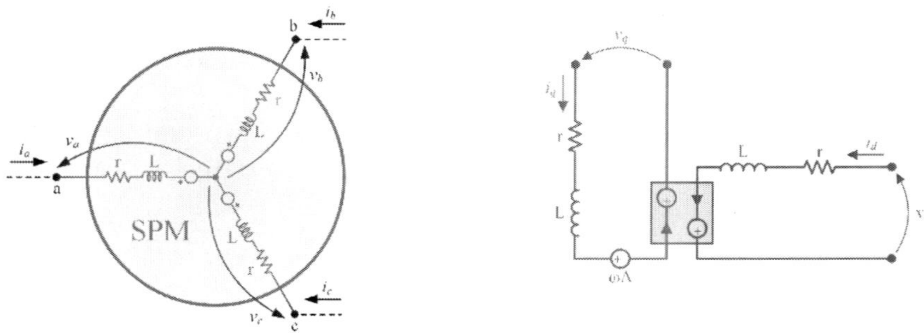

Figure 3. Schematic representation of a three-phase SPM: abc and dq equivalent circuits (on the left and on the right respectively).

Mathematical Modelling

Referring to the schematic representation of three-phase SPM depicted in Fig. 3, where *{a,b,c}* denote the phases of the stator winding, SPM electrical equations can be expressed as

$$v_n = r\,i_n + L\frac{di_n}{dt} + \frac{d\lambda_n}{dt} \quad , \qquad n \in \{a,b,c\} \tag{7}$$

in which r and L denote the phase resistance and the synchronous inductance, whereas v_n and i_n are phase voltages and currents respectively, λ_n being the magnetic flux linkages due to permanent magnets. It is worth noting that (7) is assumed in conditions of negligible magnetic anisotropy and saturation effects, as generally occurs for SPM. Thus, it is possible to define voltage and current space vectors based on their corresponding phase quantities as

$$x_{\alpha\beta} = \frac{2}{3}\left(x_a + x_b\, e^{j\frac{2}{3}\pi} + x_c\, e^{j\frac{4}{3}\pi} \right) \quad , \quad x \in \{v, i, \lambda\}\,. \tag{8}$$

Therefore, by substituting (7) in (8) and assuming each λ_n sine-shaped, the continuous-time electrical equation of SPM in the stationary αβ reference frame can be achieved as

$$v_{\alpha\beta} = r\,i_{\alpha\beta} + L\frac{di_{\alpha\beta}}{dt} + \frac{d\lambda_{\alpha\beta}}{dt} \quad , \qquad \lambda_{\alpha\beta} = \Lambda e^{j\theta} \tag{9}$$

where the constant Λ denotes the magnitude of the flux linkage vector, whereas θ is the electrical rotor position. However, moving from the stationary αβ to the synchronous dq reference frame is generally required for control system design purposes, thus (9) can be replaced by

$$v_{dq} = (r + j\omega L) i_{dq} + L \frac{di_{dq}}{dt} + j\omega\Lambda \tag{10}$$

whose circuital representation is depicted in Fig. 3, whereas ω is the electrical rotor speed and

$$x_{dq} = x_{\alpha\beta} \, e^{-j\theta} \quad , \quad x \in \{v, i, \lambda\}. \tag{11}$$

Finally, denoting by p the number of pole pairs, SPM electromagnetic torque T_e and power P_m can be expressed respectively as

$$T_e = \frac{3}{2} p\Lambda \cdot i_q \quad , \quad P_m = \frac{3}{2} \omega\Lambda \cdot i_q . \tag{12}$$

It is worth nothing that (10) is more suitable than (9) for designing SPM control algorithms, as just stated before. In fact, referring to (12), appropriate torque control can be achieved by controlling i_q directly, thus justifying the employment of the dq reference frame in place of the stationary αβ one.

Hence, in order to design an MPC algorithm for driving SPM torque to its reference value, an SPM sampled-data model is required at first. This can be achieved from (10), which, however, can be better expressed as

$$\frac{di_{idq}}{dt} = -\left(\frac{r}{L} + j\omega\right) i_{dq} + \frac{1}{L} v_{dq} - \frac{1}{L} j\omega\Lambda . \tag{13}$$

Thus, if the sampling time interval T_s is chosen sufficiently small in order to assume the rotor speed variation negligible in $[kT_s,(k+1)T_s]$, the discretization procedure can be carried out, leading to

$$i_{dq}^{(k+1)} = f_k \, i_{dq}^{(k)} + h_k \cdot \left(V_k \, \tau_{dq}^{(k)} - E_k \, \xi_{dq}^{(k)} \right) \tag{14}$$

in which V and τ denote the DC-link voltage and its corresponding pulse width vector, whereas E and ξ are the back-emf magnitude and pulse width vector respectively:

$$\xi_k = jT_s e^{j\omega_k \frac{T_s}{2}} \quad , \quad E_k = \omega_k \Lambda . \tag{15}$$

Finally, still referring to (14), the coefficients f_k and h_k are defined as

$$f_k = e^{-\left(\frac{r}{L}+j\omega_k\right)T_s} \quad , \quad h_k = \frac{1}{L} e^{-\left(\frac{r}{2L}+j\omega_k\right)T_s} . \tag{16}$$

It is worth noting that (14) is achieved on condition that the chain voltage pulses applied to SPM are symmetrically spaced with respect to the middle of the sampling time interval. In fact, in such a case, the errors occurring between the exact sampled-data model [47]and (14) can be ignored[48].

SPM Operating Constraints

SPM operation is bounded by several constraints, all of which contribute to determine its steady-state operating range, as well as its dynamic performances. Firstly referring to SPM steady-state operation, current limitation and voltage saturation constraints must be always satisfied, they being expressed respectively as

$$\left| i_{dq}^{(k+1)} \right| \leq I_{max} \tag{17}$$

$$\left| \tau_{dq}^{(k)} \right| \leq \mu \frac{T_s}{\sqrt{3}} \quad , \quad \mu \in (0,1) \tag{18}$$

where I_{max} denotes the maximum allowable current vector magnitude, whereas μ is generally assumed smaller than one in order to account for some DC-link voltage reserve. This last should be set aside in order to preserve SPM dynamic performances at any speed. Thus, still referring to steady-state operation, (14) yields

$$i_{dq}^{(k+1)} = i_{dq}^{(k)} \quad \rightarrow \quad i_{dq}^{(k+1)} = -\frac{h_k E_k}{1-f_k} \xi_{dq}^{(k)} + \frac{h_k V_k}{1-f_k} \tau_{dq}^{(k)} . \tag{19}$$

Consequently, by combining (19) with(18), voltage saturation constraint can be expressed in terms of its equivalent current constraint as

$$\left| i_{dq}^{(k+1)} - \varsigma_k \right| \leq \rho_k V_k \quad , \quad \varsigma_k = -\frac{h_k E_k}{1-f_k} \xi_{dq}^{(k)} \quad , \quad \rho_k = \mu \frac{|h_k|}{|1-f_k|} \frac{T_s}{\sqrt{3}} \tag{20}$$

Considering now torque and power limitations, these can be expressed respectively as

$$\left| T_e^{(k+1)} \right| \leq T_{e,nom} \quad , \quad \left| P_m^{(k+1)} \right| \leq P_{m,nom} \tag{21}$$

where $T_{e,nom}$ and $P_{m,nom}$ denote rated torque and power values respectively, which are proportional to each other through the rated speed ($\omega_{m,nom}$). Hence, on the basis of (12) and assuming rotor speed variations negligible over T_s, (21) can be easily expressed in terms of i_q as:

$$\left|i_q^{(k+1)}\right| \leq I_q^{(k)} \quad , \quad I_q^{(k)} = min\left\{\frac{2T_{e,nom}}{3p\Lambda}, \frac{2P_{m,nom}}{3\left|\omega_k\right|\Lambda}\right\}. \tag{22}$$

Hence, referring to the (i_d,i_q) plane, (17) and (20) state that i_{dq} must belong to the circles $\mathcal{C}_1$ and $\mathcal{C}_2$ respectively, whereas (22) require that i_{dq} lies within the plane region $\mathcal{C}_3$ bounded by two straight lines, all this shown in Fig. 4. As a result, steady-state SPM operating boundaries can be achieved for any given rotor speed value. Particularly, based on (5), the subset X_0 can be defined as:

$$X_0^{(k)} : \ \forall i_{dq}^{(k+1)} \in X_0^{(k)} \ \rightarrow \ \mathcal{C}_l^{(k)} \leq 0 \quad , \quad l = 1..3 . \tag{23}$$

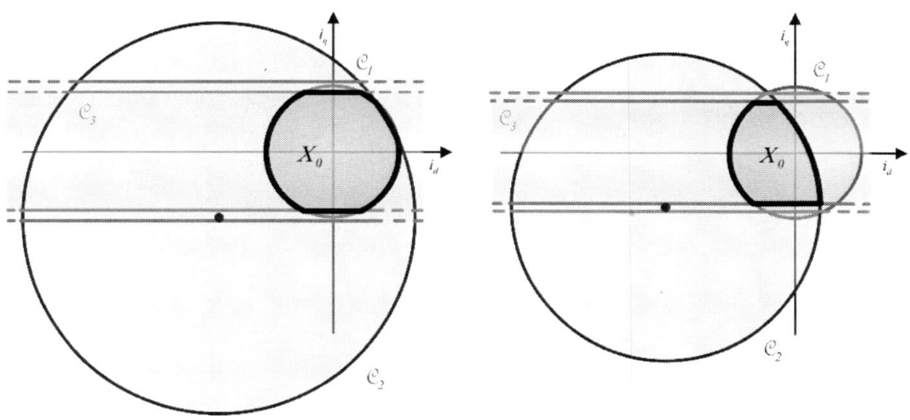

Figure 4. Steady-state SPM operating boundaries (black) on the (i_d,i_q) plane, below and above the rated speed (on the left and on the right respectively): current limitation (red), voltage saturation (blue), torque and power limitation (orange and green respectively).

It is worthy of note that, below the rated speed, power limitation does not contribute to identify SPM operating boundaries, torque limitation being predominant, as pointed out by (22). The opposite occurs above the rated speed, in correspondence of which torque limitation becomes irrelevant. In conclusion, if no DC-link voltage reserve is set aside, i.e. μ is chosen equal to one, $\mathcal{C}_2$ can be enlarged. This could be particularly suitable at very high speed operation in order to increase SPM steady-state performances further [50].

Referring now to SPM dynamic operation, (17) and (22) still occur, whereas voltage saturation constraint must be now expressed as:

$$\left|\tau_{dq}^{(k)}\right| \leq \frac{T_s}{\sqrt{3}} . \tag{24}$$

In particular, (24) is achieved from (18) by imposing μ equal to one because there is no need of setting aside any DC-Link voltage reserve over SPM dynamic operation. In addition, since also (19) cannot be assumed further, (14) yields

$$\tau_{dq}^{(k)} = \frac{1}{h_k V_k}\left(i_{dq}^{(k+1)} - f_k\, i_{dq}^{(k)}\right) + \frac{E_k}{V_k}\xi_{dq}^{(k)} . \tag{25}$$

Consequently, by substituting (25) in (24), the voltage saturation constraint over dynamic operation can be expressed in terms of its equivalent current constraint as:

$$\left|i_{dq}^{(k+1)} - c_k\right| \le r_k V_k \quad , \qquad c_k = f_k\, i_{dq}^{(k)} - h_k E_k \xi_{dq}^{(k)} \quad , \quad r_k = \left|h_k\right| \frac{T_s}{\sqrt{3}} . \tag{26}$$

Thus, still referring to the (i_d, i_q) plane, (26) states that i_{dq} must belong to the circle $\mathcal{C}_4$ shown in Fig. 5, which represents all the possible current values reachable over the $[kT_s,(k+1)T_s]$ sampling time interval. Thus, SPM dynamic operating boundaries can be achieved, as shown in Fig. 5 too. It is worth noting that SPM dynamic operation is constrained not only by $\mathcal{C}_1$, $\mathcal{C}_3$ and $\mathcal{C}_4$, but also by $\mathcal{C}_2$. In fact, if i_{dq} lies outside $\mathcal{C}_2$, SPM controllability may be impaired, leading to unpredictable current evolutions. As a result, the dynamic subset $\tilde{X}_0$ can be defined as:

$$\tilde{X}_0^{(k)} \subseteq X_0^{(k)} : \quad \forall i_{dq}^{(k+1)} \in \tilde{X}_0^{(k)} \quad \rightarrow \quad \mathcal{C}_l^{(k)} \le 0 \quad , \quad l = 1..4 . \tag{27}$$

In conclusion, it is also worthy of note that voltage saturation constraint affects SPM steady-state and dynamic operation differently, leading to different SPM operating boundaries. This suggests the need of managing steady-state and dynamic operating constraints separately in order to design an optimal MPC algorithm, as detailed in the following.

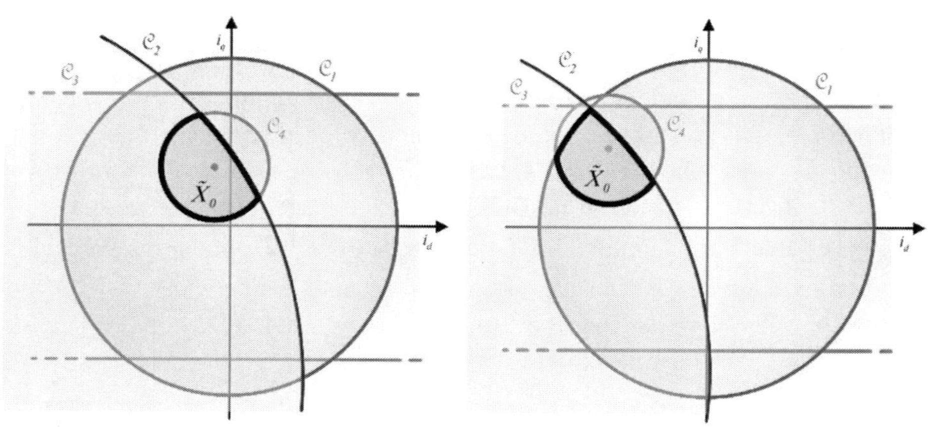

Figure 5. SPM dynamic boundaries on the (i_d, i_q) plane over dynamic and steady-state operation (on the left and on the right respectively): current limitation (red), voltage saturation (blue, cyan) and power limitation (green).

MPC Design

In order to design an MPC algorithm for SPM torque control, steady-state and dynamic operation should be managed separately, as stated before. In fact, for a given reference torque value ($T_e^{(ref)}$), the corresponding reference current vector (i_{dq}*) should be determined at first in order to optimize SPM steady-state performances. Thus, the following steady-state objective functions are introduced:

$$\varphi_1^{(k)} = \left| T_e^{(ref)} - \frac{3}{2} p\Lambda \cdot i_q^* \right|_k \quad , \quad \varphi_2^{(k)} = \frac{3}{2} r \cdot \left| i_{dq}^* \right|_k^2 \tag{28}$$

whereas the following constraints must be satisfied:

$$\mathcal{C}_1^{(k)} : \left| i_{dq}^{(k+1)} \right| \le I_{max}, \quad \mathcal{C}_2^{(k)} : \left| i_{dq}^{(k+1)} - \varsigma_k \right| \le \rho_k V_k, \quad \mathcal{C}_3^{(k)} : \left| i_q^{(k+1)} \right| \le I_q^{(k)}. \tag{29}$$

In particular, φ_1 entails that SPM electromagnetic torque should be driven to its reference value by means of i_q*, whereas φ_2 imposes i_d* constant at zero in order to minimize Joule losses. It is worth noting that φ_1 is properly prioritized with respect to φ_2 in order to achieve an optimal SPM torque control. In fact, at high speed operation, d current components different from zero could be needed in order to achieve the reference torque value, leading to flux-weakening operation. This can be successfully accomplished by prioritizing φ_1, whereas it can be much less guaranteed by a conventional MPC cost function, like Φ defined by (3).

Once the reference current vector has been set, it should be tracked as well as possible over SPM dynamic operation. This goal can be successfully achieved by minimizing the following dynamic objective functions:

$$\varphi_3^{(k)} = \left| i_m^* \right|_k - i_m^{(k+1)} \Big| \quad , \quad \varphi_4^{(k)} = \left| i_n^* \right|_k - i_n^{(k+1)} \Big| \quad , \quad mn \in \{dq, qd\} \tag{30}$$

and by satisfying (29) and the following additional dynamic constraint simultaneously:

$$\mathcal{C}_4^{(k)} : \left| i_{dq}^{(k+1)} - c_k \right| \le r_k V_k . \tag{31}$$

Particularly, m alternatively denotes d or q over SPM flux-weakening or flux-strengthening operation respectively, the opposite occurring for n. Such a choice allows the achievement of the reference current vector by satisfying (29) automatically; consequently, MPC algorithm implementation can be eased without resorting to any SPM model and/or constraint approximation.

Summarizing, for a given $T_e^{(ref)}$, the most suitable i_{dq}* for minimizing φ_1 and, eventually, φ_2, is synthesized at first. Subsequently, such a reference current value is tracked as well as possible by selecting the $i_{dq}^{(k+1)}$ value in order to minimize φ_3 and φ_4 in descending order of priority. Finally, once $i_{dq}^{(k+1)}$ is achieved, the voltage pulse width vector $\tau_{dq}^{(k)}$ can be easily computed by means of (25), all steady-state and dynamic constraints being satisfied. The mathematical procedure for designing such an MPC algorithm is thus detailed in the following two sub-sections, which refers to the management of SPM steady-state and dynamic operation respectively, whereas an overview of the MPC algorithm as a whole is reported in the last sub-section.

Steady-State Operation Management

In order to ease SPM constraint management over steady-state operation, the following variables are introduced at first:

$$\Psi_k = \frac{\left| i_{dq}^{(k+1)} - \varsigma_k \right|^2}{\rho_k^2} \quad , \quad \Phi_k = \left| i_{dq}^{(k+1)} \right|^2 \quad , \quad \Gamma_k = \frac{2T_e^{(k+1)}}{3p\Lambda} . \tag{32}$$

In particular, referring to(29), Ψ and Φ account for square magnitudes of voltage and current vectors respectively, whereas Γ denotes the torque current, as well highlighted by (22). Thus, by substituting (32) in (29), SPM operating constraints can be easily expressed as

$$\mathcal{C}_1^{(k)}: \ \sqrt{\Phi_k} \le I_{max}, \qquad \mathcal{C}_2^{(k)}: \ \sqrt{\Psi_k} \le V_k, \qquad \mathcal{C}_3^{(k)}: \ \left|\Gamma_k\right| \le I_q^{(k)} . \tag{33}$$

As a result, each SPM steady-state constraint affect only one of $\{\Psi,\Phi,\Gamma\}$, thus their management can be performed more easily than referring to i_{dq}. However, it is worth noting that $\{\Psi,\Phi,\Gamma\}$ are not independent; in fact, the combination of (32) to each other leads to

$$\Xi_k^T \, Q_k \, \Xi_k = 0 \tag{34}$$

where the vector Ξ_k and the matrix Q_k are defined as follows:

$$\Xi_k = \begin{bmatrix} \Psi \\ \Phi \\ \Gamma \\ 1 \end{bmatrix}_k \quad , \quad Q_k = \begin{bmatrix} \rho^4 & -\rho^2 & 2\rho^2 \varsigma_q & -\rho^2 |\varsigma|^2 \\ -\rho^2 & 1 & -2\varsigma_q & -\varsigma_d^2 + \varsigma_q^2 \\ 2\rho^2 \varsigma_q & -2\varsigma_q & 4|\varsigma|^2 & -2\varsigma_q |\varsigma|^2 \\ -\rho^2 |\varsigma|^2 & -\varsigma_d^2 + \varsigma_q^2 & -2\varsigma_q |\varsigma|^2 & |\varsigma|^4 \end{bmatrix}_k . \tag{35}$$

In particular, (34) states that the vector $[\Psi,\Phi,\Gamma]$ must lie within an elliptic paraboloid, whose intersections with planes parallel to (Ψ,Φ) identify constant-torque parabolas. Hence, in order to minimize φ_1 and, possibly, φ_2, a Maximum Torque per Ampere (MTPA) control strategy should be employed, which consists in minimizing Φ for a given Γ. Consequently, based on(34), such an operating condition yields:

$$\Psi_k^{(MTPA)} = \frac{1}{\rho_k^2} \left| \varsigma_{d,k} + j\left(\Gamma_k - \varsigma_{q,k}\right) \right|^2 \quad , \quad \Phi_k^{(MTPA)} = \Gamma_k^2 . \tag{36}$$

However, (36) does not always comply with (33), especially at high speed operation; thus, assuming that $\Gamma^{(ref)}$ is chosen in accordance with $\mathcal{C}_3$, different cases should be considered, in accordance with the rotor speed value. In particular, when the rotor speed is below its rated value, like the case depicted in Fig. 6, (36) complies with both current limitation and voltage saturation constraints, leading to:

$$\Psi_k^* = \frac{1}{\rho_k^2}\left| \varsigma_{d,k} + j\left(\Gamma_k^* - \varsigma_{q,k}\right)\right|^2 \quad , \quad \Phi_k^* = \left(\Gamma_k^*\right)^2 \quad , \quad \Gamma_k^* = \Gamma_k^{(ref)} . \tag{37}$$

As a result, optimal SPM steady-state operation occurs on the point A shown in Fig. 6, in correspondence of which φ_1 and φ_2 are appropriately minimized in descending order of priority as

$$\varphi_1^{(k)} = 0 \quad , \quad \varphi_2^{(k)} = \frac{3}{2} r \cdot \left(\Gamma_k^*\right)^2 . \tag{38}$$

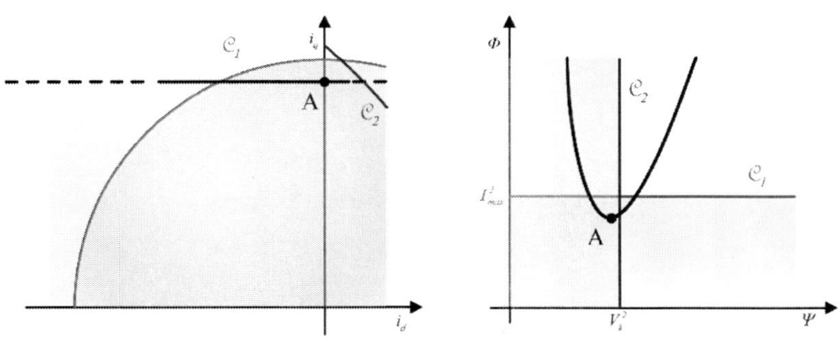

Figure 6. SPM steady-state operation below the rated speed on the (i_d, i_q) and (Ψ, Φ) planes (on the right and on the left respectively): constant-torque locus (black), current limitation (red) and voltage saturation (blue).

Differently, above the rated speed, (36) does not generally comply with voltage saturation constraint, as shown in Fig. 7. Therefore, Ψ, Φ and Γ have to be chosen in accordance with the following relationships:

$$\Psi_k^* = V_k^2, \quad \Phi_k^* = \left| \varsigma_d^{(k)} + \sqrt{\rho_k^2 \Psi_k^* - \left(\Gamma_k^* - \varsigma_q^{(k)}\right)^2} + j\Gamma_k^* \right|^2 , \quad \Gamma_k^* = \Gamma_k^{(ref)} . \tag{39}$$

As a result, SPM should operate in correspondence of the point B shown in Fig. 7; in fact, such operating condition assures that φ_1 is still held at zero and that φ_2 is minimized as much as possible:

$$\varphi_1^{(k)} = 0 \quad , \quad \varphi_2^{(k)} = \frac{3}{2} r \cdot \left(\left(\Gamma_k^*\right)^2 + \left(\varsigma_d^{(k)} + \sqrt{\rho_k^2 \Psi_k^* - \left(\Gamma_k^* - \varsigma_q^{(k)}\right)^2} \right)^2 \right) . \tag{40}$$

The comparison between (40) and (38) reveals that, above the rated speed, φ_1 can still be minimized, whereas φ_2 is increased due to flux-weakening operation. However, SPM constant-power operation can be assured on condition that current limitation constraint is fulfilled.

As the rotor speed increases, (39) does not comply with current limitation constraint further, like the case depicted in Fig. 8. Consequently, both Ψ and Φ are imposed equal to their corresponding maximum values, whereas Γ has to be computed by means of (34):

$$\Psi_k^* = V_k^2 \quad , \quad \Phi_k^* = I_{max}^2 \quad , \quad \Gamma_k^* = \Gamma_k\left(\Psi_k^*, \Phi_k^*, \Gamma_k^{(ref)}\right). \tag{41}$$

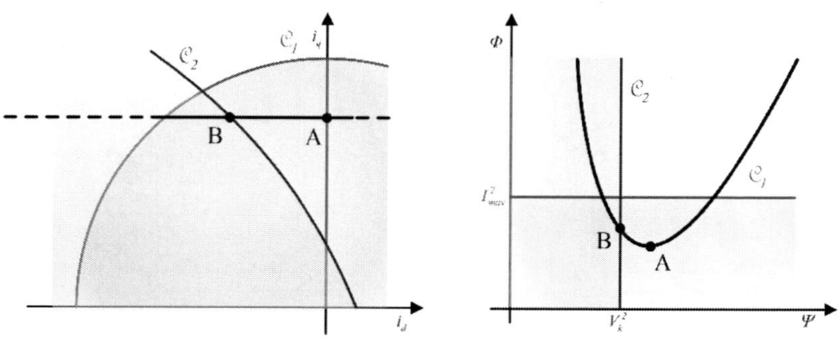

Figure 7. SPM steady-state operation above the rated speed on (i_d, i_q) and (Ψ, Φ) planes (on the right and on the left respectively): constant-torque locus (black), current limitation (red) and voltage saturation (blue).

Over such an operating condition, which corresponds to the point C shown in Fig. 8, the reference torque value cannot be achieved further, thus reduced-power operation occurs.

The equivalent block control scheme of the proposed MPC approach for SPM steady-state operation management is depicted in Fig. 9. It can be seen that, once Ψ, Φ and Γ have been determined, the corresponding reference current vector can be achieved from (32) as

$$i_{dq}^*\Big|_k = \left(\gamma_k^* \pm j\sqrt{\frac{\Phi_k^*}{|\varsigma_k|^2} - \left(\gamma_k^*\right)^2}\right) \cdot \varsigma_k \quad , \quad \gamma_k^* = \frac{1}{2}\left(1 + \frac{\Phi_k^* - \rho_k^2 \Psi_k^*}{|\varsigma_k|^2}\right). \tag{42}$$

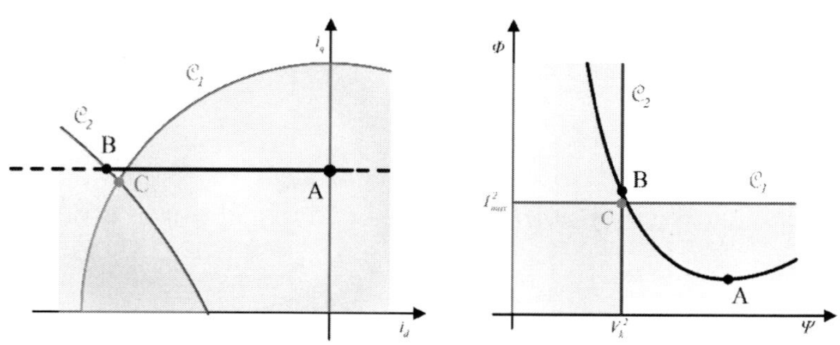

Figure 8. SPM steady-state operation at high-speed on the (i_d, i_q) and (Ψ, Φ) planes (on the right and on the left respectively): constant-torque locus (black), current limitation (red) and voltage saturation (blue).

Dynamic Operation Management

Once the most suitable reference current vector has been synthesized, the MPC has to guarantee its tracking over each $[kT_s,(k+1)T_s]$ sampling time interval. In particular, the voltage pulse width vector has to be determined in accordance with (25), which is reported in the following for convenience purpose:

$$\tau_{dq}^{(k)} = \frac{1}{h_k V_k}\left(i_{dq}^{(k+1)} - f_k\, i_{dq}^{(k)}\right) + \frac{E_k}{V_k}\xi_{dq}^{(k)} \tag{25}$$

Consequently, based on (25), τ_{dq} can be computed by a suitable choice of $i_{dq}^{(k+1)}$. However, such a choice must be performed not only in accordance with the reference current vector, but also in order to satisfy all SPM operating constraints, i.e. $\mathcal{C}_1$ through $\mathcal{C}_4$. Thus, different situations can occur, depending on reference and actual i_{dq} values. In fact, if i_{dq}* satisfies $\mathcal{C}_4$, it means that such a reference value can be achieved at $(k+1)T_s$ because it already complies with $\mathcal{C}_1$ through $\mathcal{C}_3$, as shown in Fig. 10. As a result, τ_{dq} can be computed directly as

$$i_{dq}^{(k+1)} = i_{dq}^{*}\Big|_k \quad \rightarrow \quad \tau_{dq}^{(k)} = \frac{1}{h_k V_k}\left(i_{dq}^{*}\Big|_k - f_k\, i_{dq}^{(k)}\right) + \frac{E_k}{V_k}\xi_{dq}^{(k)} \, . \tag{43}$$

In such cases, which generally occur at the end of transient responses and over steady-state operation, the dynamic objective functions φ_3 and φ_4, already defined by (30), are both minimized:

$$\varphi_3^{(k)} = 0 \quad , \quad \varphi_4^{(k)} = 0 \, . \tag{44}$$

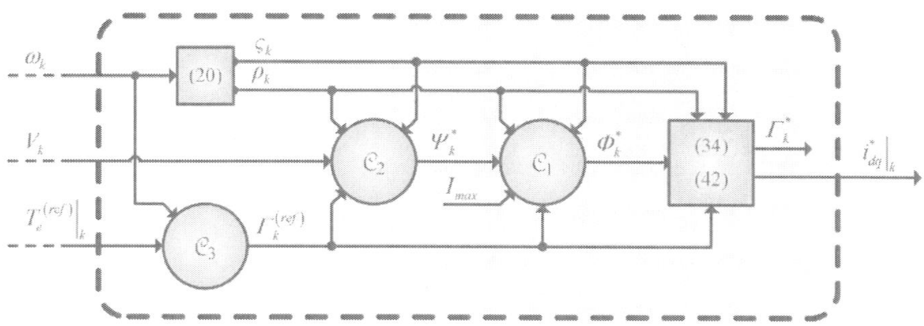

Figure 9. Equivalent block control scheme of the proposed MPC for SPM steady-state operation management.

Otherwise, if i_{dq}* does not comply with $\mathcal{C}_4$, as generally occurs over dynamic operation, an approaching strategy to the reference current vector must be employed. In these cases, $i_{dq}^{(k+1)}$ can be chosen in order to minimize the dynamic objective functions φ_3 and φ_4; this guarantees the approaching to the reference current vector, by satisfying $\mathcal{C}_1$ through $\mathcal{C}_3$ at the same time. In particular, referring to (30), the approaching to i_d* has to be prioritized compared to i_q* over flux-weakening operation in order to preserve SPM controllability, as shown in Fig. 10. Consequently, (30) becomes:

$$\Re\left\{ i^*_{dq}\Big|_k - c_k \right\} \le 0 \quad \rightarrow \quad mn = dq \rightarrow \quad \begin{cases} \varphi_3^{(k)} = \left| i^*_d\Big|_k - i_d^{(k+1)} \right| \\ \varphi_4^{(k)} = \left| i^*_q\Big|_k - i_q^{(k+1)} \right| \end{cases} \tag{45}$$

Such a priority order is due to the fact that, over flux-weakening operation, d current components different from zero are injected in order to satisfy steady-state voltage saturation constraint ($\mathcal{C}_2$). Consequently, the tracking of i_d* should be prioritized compared to i_q*, whose achievement strictly depends on that of i_d*. As a result, $i_{dq}{}^{(k+1)}$ is chosen in accordance with $\mathcal{C}_4$ as

$$\begin{aligned} i_d^{(k+1)} &= c_d^{(k)} + max\left\{ -r_k V_k, min\left\{ i_d^* - c_d^{(k)}, r_k V_k \right\} \right\} \\ i_q^{(k+1)} &= c_q^{(k)} + max\left\{ -\delta_q^{(k+1)}, min\left\{ i_q^* - c_q^{(k)}, \delta_q^{(k+1)} \right\} \right\} \\ \delta_q^{(k+1)} &= \sqrt{\left(r_k V_k\right)^2 - \left(i_d^{(k+1)} - c_d^{(k)}\right)^2}. \end{aligned} \tag{46}$$

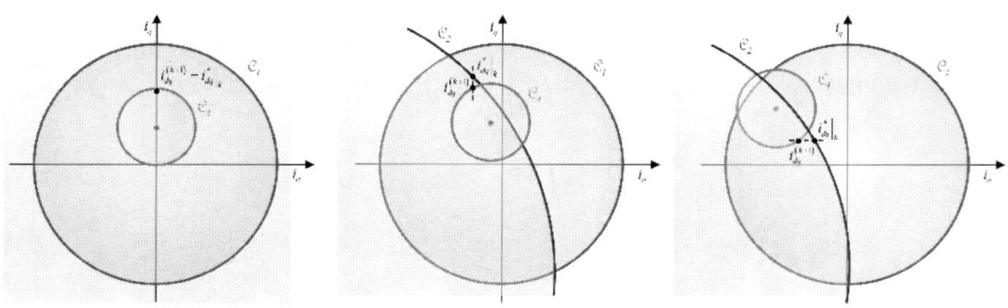

Figure 10. SPM dynamic operation on the (i_d,i_q) plane at the end of an SPM start-up (on the left), over flux-weakening operation (in the middle) and over flux-strengthening operation (on the right).

Contrariwise, over flux-strengthening operation, the approaching to i_q* should be prioritized over i_d*, leading to

$$\Re\left\{ i^*_{dq}\Big|_k - c_k \right\} > 0 \quad \rightarrow \quad mn = qd \quad \rightarrow \quad \begin{cases} \varphi_3^{(k)} = \left| i^*_q\Big|_k - i_q^{(k+1)} \right| \\ \varphi_4^{(k)} = \left| i^*_d\Big|_k - i_d^{(k+1)} \right| \end{cases} \tag{47}$$

In fact, such a choice successfully preserves SPM controllability over dynamic operation, as well shown in Fig. 10. Consequently, $i_{dq}{}^{(k+1)}$ has to be chosen in accordance with $\mathcal{C}_4$ as

$$i_d^{(k+1)} = c_d^{(k)} + max\left\{-\delta_d^{(k+1)}, min\left\{i_d^* - c_d^{(k)}, \delta_d^{(k+1)}\right\}\right\}$$
$$i_q^{(k+1)} = c_q^{(k)} + max\left\{-r_k V_k, min\left\{i_q^* - c_q^{(k)}, r_k V_k\right\}\right\} \quad (48)$$
$$\delta_d^{(k+1)} = \sqrt{\left(r_k V_k\right)^2 - \left(i_q^{(k+1)} - c_q^{(k)}\right)^2}.$$

In conclusion, the equivalent block control scheme of the proposed MPC approach for SPM dynamic operation management is depicted in Fig. 11. In particular, by substituting either (46) or (48) in (25), it is possible to synthesize the voltage pulse width vector to be applied in $[kT_s,(k+1)T_s]$. In this context, it is worth noting that, as soon as $i_{dq}*$ complies with $\mathcal{C}_4$, both (46) and (48) coincide with (43). Consequently, the employment of (43) is not directly required for implementing the proposed MPC.

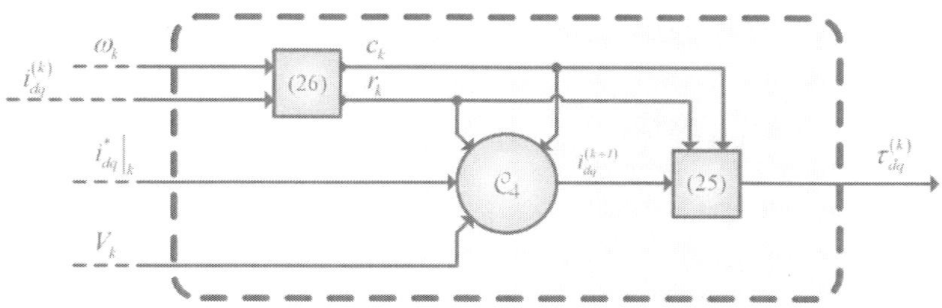

Figure 11. Equivalent block control scheme of the proposed MPC for SPM dynamic operation management.

MPC Overview

The MPC algorithm as a whole can be summarized as in Fig. 12. Particularly, it can be seen that it starts as soon as DC-link voltage, actual phase currents, rotor speed and position become available by means of appropriate sensors. Thus, MPC algorithm focuses on SPM steady-state operation management at first: in particular, for a given $T_e^{(ref)}$, $\Psi*$, $\Phi*$ and $\Gamma*$ are determined in accordance with $\mathcal{C}_1$ through $\mathcal{C}_3$. As a result, the most suitable reference current vector $i_{dq}*$ is achieved. Meanwhile, dq current space vector is computed based on phase currents and rotor position. Then, SPM dynamic operation management is carried out based on $i_{dq}*$ and $\mathcal{C}_4$, which allows the computation of the voltage pulse width vector τ_{dq}. This last is then successfully employed in order to synthesize appropriate Pulse Width Modulation (PWM) signals, by means of which the SPM can be successfully driven in accordance with its reference torque and current profiles.

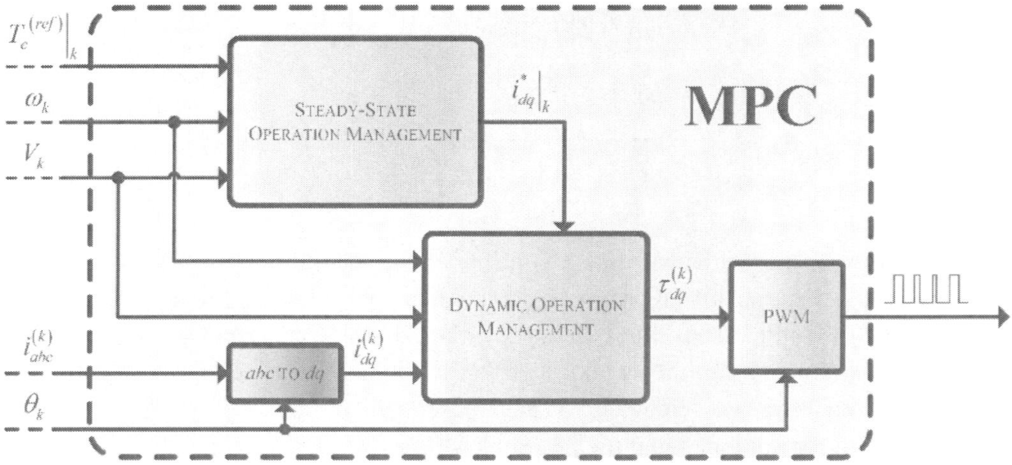

Figure 12. The overall MPC block control scheme.

Simulations and Experiments

In order to verify the effectiveness of the novel MPC algorithm considered in this chapter, a simulation study is carried out in the Matlab Simulink environment at first. Subsequently, several experimental tests have been performed, which also regard the comparison between the performances achievable by means of the novel MPC algorithm and those obtainable by a conventional PI-based control system, i.e. the well-known voltage follower based on a PI voltage compensator (VF-PI) [43]-[46].

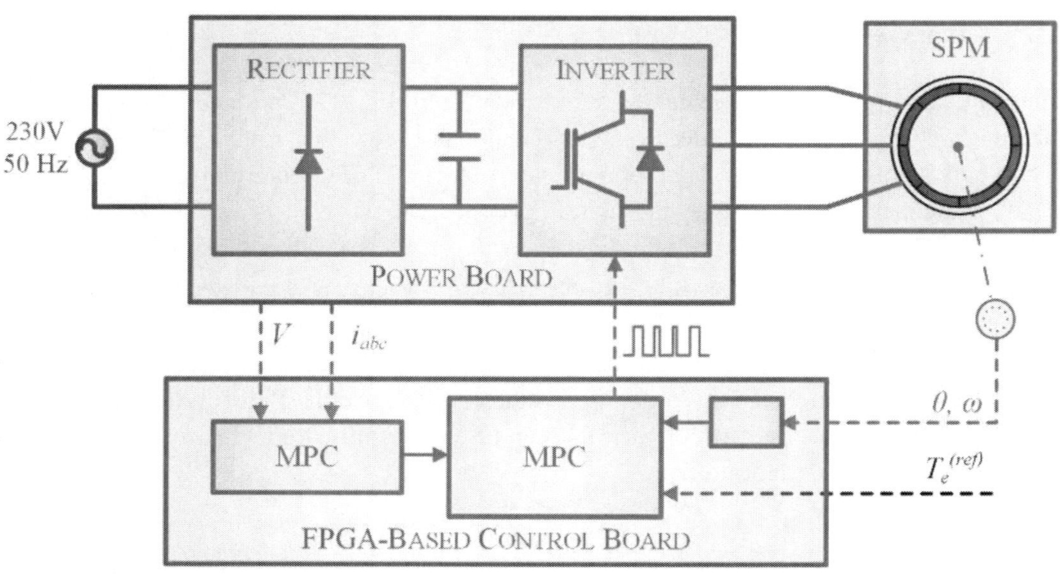

Figure 13. Simulation and experimental setup.

Both simulations and experiments have been performed with reference to the setup shown in Fig. 13; in particular, an SPM is considered, whose parameters and rated values are reported in Table I. The SPM is fed through a power board, which consists mainly of a diode bridge rectifier and a three-phase inverter; this last is equipped with an IGBT smart power module (600 V, 15 A), the DC link voltage being about 325 V. In this context, in order to properly account for inverter un-idealities, simulations are carried out by assuming the forward voltage drops of IGBTs and freewheeling diodes equal to 0.6 V and 0.8 V respectively, their equivalent conduction resistance having been set to 0.1 Ω. The inverter supplies the SPM in accordance with appropriate PWM signals, whose dead-time and minimum pulse width have been set to 1 μs and 0.8 μs respectively. Both simulations and experiments are characterized by a switching frequency of about 10 kHz and a sampling time interval of 102.4 μs.

Table I. SPM Parameters and Rated Values

	r	L	p	Λ	$T_{e,n}$	$\omega_{m,n}$	$P_{m,n}$	I_n
Value	12.1	23.4	4	0.1062	1.12	2500	0.29	1.76
Unit	Ω	mH	-	V·s	Nm	rpm	kW	A

Table II. Cyclone IV EP4CE55F23C7 specifications

Resources	LEs	I/O Pins	Memory (Kbits)	PLLs
Overall	55856	325	2396.160	4
Consumed (MPC)	41605 (74%)	24 (7%)	458.752 (29%)	1 (25%)

Table III. MPC LEs/Time Consumptions

	abc to dq	Steady-State Op.	Dynamic Op.	PWM	MPC (Total)
LEs	2272 (5.7%)	23947 (60.1%)	10818 (27.1%)	2811 (7.1%)	39848 (100%)
Time (μs)	0.16	1.60	0.64	0.32	2.36

The MPC and the VF-PI algorithms have both been implemented on an Altera Cyclone IV FPGA device, whose main specifications are shown in Table II. In particular, this last also accounts for the resources consumed by the proposed MPC implementation, including appropriate acquisition software routines for current, voltage and position sensors. In this context, phase currents and DC-Link voltage signals coming from the power board are acquired by means of FPGA input pins, some of which are also employed for the acquisition of rotor position through a high resolution digital encoder (23 bits). All these input signals are properly managed in order to obtain per-unit quantities, particularly a fixed-point format of 15 bits is employed, the integer part consisting in the most significant bit only. Regarding the memory usage, it is due to the trigonometric functions look-up table, whereas the PLL allows the generation of several clock signals from that provided by the main board (50 MHz).

The logic elements (LEs) and time consumptions of each block of the MPC scheme depicted in Fig. 12 are summarized in Table III. It can be noticed that steady-state operation management uses the most significant part of the overall MPC consumed resources, in terms of both LEs and execution time. This is due to the need of an accurate management of steady-state operating constraints, as shown in Fig. 9, which allows the optimization of SPM performances within its operating boundaries. Much less resources are required by dynamic operation management, as also detectable by Fig. 11; in fact, in this case, the novel MPC approach leads to automatically satisfy $\mathcal{C}_1$ through $\mathcal{C}_3$, thus only $\mathcal{C}_4$ has to be considered. However, it is worth noting that the overall resources consumed by the MPC algorithm are quite limited, both in terms of LEs and execution times, highlighting the effectiveness of the proposed implementation procedure. In particular, the synchronous design approach has been employed, thus each block has been coded in VHDL directly and consists of several math operations (including divisions, multiplications and square roots). These are parallelized as much as possible in order to minimize the MPC algorithm execution time, as shown in Fig. 12 and Table III. Consequently, the PWM signals can be immediately applied in the same sampling time interval, resulting in a real-time implementation of the MPC algorithm.

Simulation and Experimental Results

Simulation and experimental results firstly focus on SPM steady-state operation management performed by the novel MPC algorithm. Thus, reference can be made to a start-up of the drive, as shown in Fig. 14 through Fig. 16. In particular, starting from rest, a step-reference torque value is imposed, equal to its rated one. Consequently, the rotor speed increases until steady-state operation is reached. Firstly referring to Fig. 14, it can be seen that, below the rated speed, i_q is driven to its rated value, whereas i_d is held constant at zero. As a result, SPM electromagnetic torque successfully tracks its reference value, whereas SPM output power increases with the rotor speed, as highlighted in Fig. 15. It is worth noting that, in this case, MTPA operation can be successfully achieved without resorting to flux-weakening operations. Consequently, referring to Fig. 16, Φ is constant at its rated value, whereas Ψ increases with the rotor speed up to its maximum value. Over such an operating conditions, the proposed MPC algorithm is thus able to nullify φ_1 and minimize φ_2, as stated by (38).

Subsequently, flux-weakening operation occurs due to the increased rotor speed. Thus, as soon as the rotor speed overcomes its rated value, i_q* is driven in accordance with $\mathcal{C}_3$ for assuring SPM constant-power operation, as highlighted in both Fig. 14 and Fig. 15. Whereas i_d* arises just in order to hold Ψ constant at its maximum value, as shown in Fig. 16. As a result, φ_1 can still be nullified, whereas φ_2 initially decreases and subsequently increases with the rotor speed in accordance with the Φ evolution depicted in Fig. 16. However, as soon as Φ reaches its maximum value, constant-power operation cannot be assured further, as clearly shown in Fig. 15. Over such an operating condition, reduced-power operation occurs, both Ψ and Φ being held constant at their corresponding maximum values, as shown in Fig. 16 and stated by (41). Consequently, φ_1 increases with the rotor speed, φ_2 being constant at its maximum value. Summarizing, it can be stated that simulation and experimental results reveal the effectiveness of the MPC algorithm in terms of SPM steady-state operation management. In addition, it is also worthy of note the very good agreement between simulations and experiments, which both highlight the fact that reference and actual current evolutions are almost superimposed, as shown in Fig. 14.

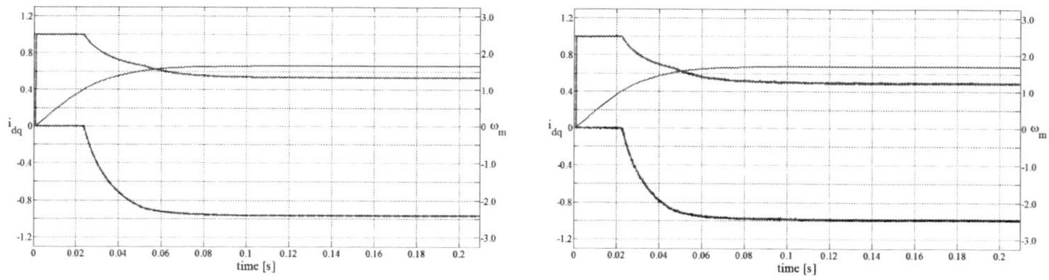

Figure 14. The dq current component and rotor speed evolutions achieved in simulations (on the left) and experiments (on the right): i_d^* (pink), i_q^* (cyan), i_d (brown), i_q (turquoise) and ω_m (green).

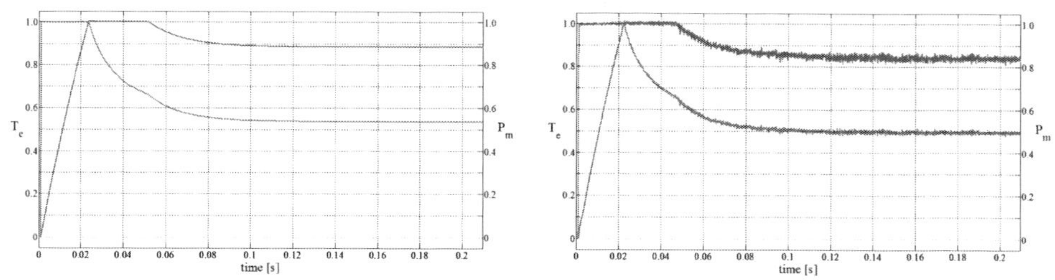

Figure 15. The electromagnetic torque (orange) and output power (green) achieved in simulations (on the left) and derived from experimental data (on the right).

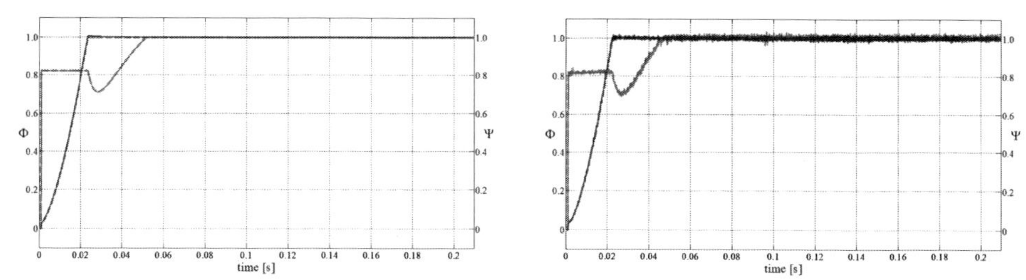

Figure 16. The square magnitudes of the voltage and current vectors achieved in simulations (on the left) and experiments (on the right): $\boldsymbol{\Phi}^*$ (pink), $\boldsymbol{\Phi}$ (red), $\boldsymbol{\Psi}^*$ (turquoise) and $\boldsymbol{\Psi}$ (blue).

Therefore, other two experimental tests have been carried out in order to highlight the effectiveness of the novel MPC algorithm in terms of SPM dynamic operation management. In this context, the employment of a conventional VF-PI control system is also considered for comparison purposes. The first experimental test has been performed by imposing a reference torque equal to the rated one for both MPC and VF-PI algorithms. Whereas the load torque is increased compared to the previous case in order to achieve a steady-state speed value within the constant-power region. The corresponding results are thus reported in Fig. 17. Focusing on the start-up of the drive, it can be noticed that the MPC algorithm guarantees better dynamic performances compared to VF-PI: in fact, i_q* is achieved in just three sampling time intervals against the eight required by VF-PI. In addition, both i_d* and i_d are successfully held at zero in accordance with (45), which entails prioritizing the achievement of i_d* rather than i_q*. This does not occur by means of VF-PI due to the outer voltage loop. In fact, this last detects a voltage deficit over SPM dynamic operation, so it inappropriately starts flux-weakening operation, leading to unsuitable i_d*. Considering now the overall current and rotor speed evolutions achieved by MPC and VF-PI, it can be seen that they are very similar to each other. However, although both i_q*and i_q are almost the same in both cases, the reference and actual d current profiles achieved by the MPC algorithm are slightly lower than those obtained by VF-PI, all converging to the same steady-state value. This means that stronger and faster flux-weakening operation is provided by MPC than by VF-PI, revealing better dynamic performances of the former.

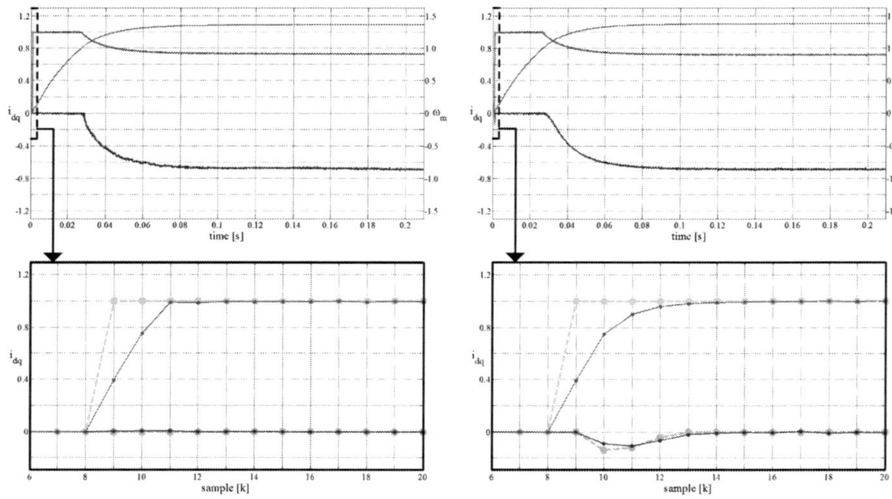

Figure 17. The dq current component and rotor speed evolutions achieved at an SPM start-up by MPC (left) and VF-PI (right): i_d^* (pink), i_q^* (cyan), i_d (brown), i_q (turquoise) and ω_m (green).

Subsequently, a second experimental test regarding an i_q inversion has been performed in order to further highlight SPM dynamic performances achievable by means of the proposed MPC algorithm. Hence, starting from rest, a step reference torque value is imposed, slightly lower than the rated one. Then, as soon as steady-state operation occurs, reference torque value is suddenly inverted, leading to the current and rotor speed evolutions depicted in Fig. 18. It can be seen that, over the i_q inversion, the MPC algorithm prioritizes the achievement of i_q* rather than i_d* due to SPM flux-strengthening operation, as stated by (47). Consequently, i_q* is quickly achieved at first, then i_d is successfully driven to i_d*. Poorer performances are achieved by means of VF-PI, i.e. a slower i_q inversion is performed, whereas i_d* even decreases due to the voltage deficit detected by the outer voltage loop, then it very slowly converges to zero, as well shown in Fig. 18.

In conclusion, simulation and experimental results show the effectiveness of this MPC approach, which allows the optimization of SPM performances over both steady-state and dynamic operation. In addition, all input and output constraints are always fulfilled, in particular SPM controllability is preserved appropriately over any operating condition.

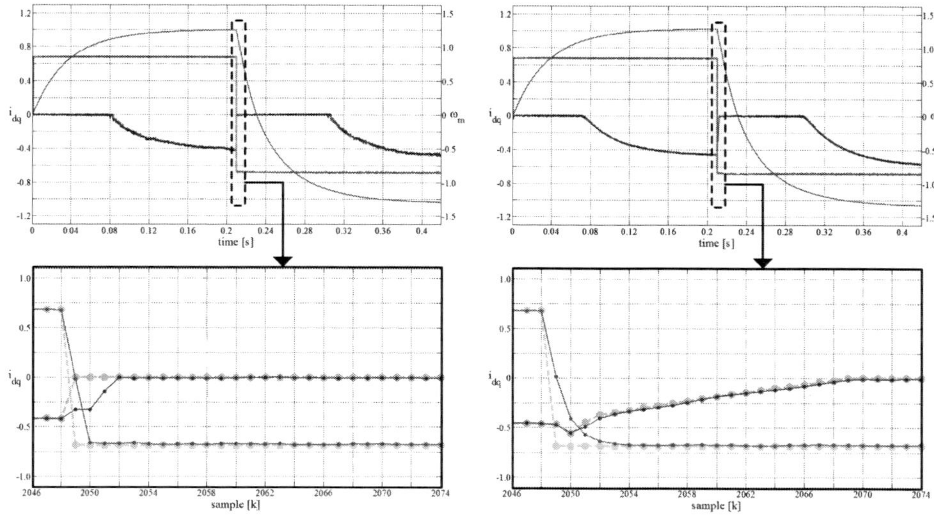

Figure 18. The dq current component and rotor speed evolutions achieved over an SPM i_q inversion by MPC (left) and VF-PI (right): i_d^* (pink), i_q^* (cyan), i_d (brown), i_q (turquoise) and ω_m (green).

Conclusion

This chapter has addressed the problem arising from input and output constraint management in designing Model Predictive Control (MPC) algorithms. In particular, conventional MPC cost function may be unsuitable for fully exploiting system performances over both dynamic and steady-state operation, as well as for satisfying all input and output constraints. An alternative solution has been thus considered, which consists of performing accurate and separate managements of steady-state and dynamic operating constraints at first. This allows the achievement of both steady-state and dynamic operating boundaries, within which appropriate objective functions can be minimized in descending order of priority. Such an alternative MPC design procedure aims to guarantee a better exploitation of system performances compared to that achievable by minimizing a conventional MPC cost function.

This novel MPC approach is validated by referring to a case study, namely a Surface-Mounted Permanent Magnet Synchronous Machine (SPM) driven by a three-phase inverter. In particular, SPM mathematical modelling and its input and output constraints have been introduced at first. Then, the MPC design procedure has been reported in detail and extensively discussed. The need of managing SPM steady-state and dynamic operation separately has been clearly highlighted, as well as the advantages of introducing different objective functions to be minimized in descending order of priority. In conclusion, simulation and experimental results are also reported, which both point out the effectiveness of the novel MPC approach, also in comparison with a conventional PI-based control system.

References

[1] R. Krishnan, *Permanent Magnet Synchronous and Brushless DC Motor Drives,* CRC Press, NW, 2010.

[2] K.P. Gokhale, A. Kawamura, R.G. Hoft, "Dead Beat Microprocessor Control of PWM Inverter for Sinusoidal Output Waveform Synthesis", *IEEE Trans. Ind. Appl.,* vol. IA-23, no. 5, pp. 901-910, Sept. 1987.

[3] A. Kawamura, R. Chuarayapratip, T. Haneyoshi, "Deadbeat control of PWM inverter with modified pulse patterns for uninterruptible power supply", *IEEE Trans. Ind. Electron.,* vol. 35, no. 2, pp. 295-300, May 1988.

[4] L. Ben-Brahim, A. Kawamura, "Digital Current Regulation of Field-Oriented Controlled Induction Motor Based on Predictive Flux Observer", *IEEE Trans. Ind. Appl.*, vol. 27, no. 5, pp. 956-961, Sept./Oct. 1991.

[5] Y. A. R. I. Mohamed, E. F. El-Saadany, "An improved deadbeat current control scheme with a novel adaptive self-tuning load model for a three-phase PWM voltage-source inverter", *IEEE Trans. Ind. Electron.*, vol. 54, no. 2, pp. 747-759, Apr. 2007.

[6] K. Sakata, H. Fujimoto, "Perfect tracking control of servo motor based on precise model with PWM hold and current loop", *in Proc. Power Conversion Conference (PCC)*, Nagoya (Japan), Apr. 2-5, 2007, pp. 1612-1617.

[7] X. Lin-Shi, F. Morel, A.M. Llor, B. Allard, J.-M. Retif, "Implementation of hybrid control for motor drives", *IEEE Trans. Ind. Electron.*, vol. 54, no. 4, pp. 1946-1952, Aug. 2007.

[8] P. Cortes, M. P. Kazmierkowski, R. M. Kennel, D. E. Quevedo, J. Rodriguez, "Predictive control in power electronics and drives", *IEEE Trans. Ind. Electron.*, vol. 55, no. 12, pp. 4312-4324, Dec. 2008.

[9] S. Kouro, P. Cortes, R. Vargas, U. Ammann, J. Rodriguez, "Model predictive control - A simple and powerful method to control power converters", *IEEE Trans. Ind. Electron.*, vol. 56, no. 6, pp. 1826-1838, Jun. 2009.

[10] H. Miranda, P. Cortes, J.I. Yuz, J. Rodriguez, "Predictive Torque Control of Induction Machines based on State-Space Models", *IEEE Trans. Ind. Electron.*, vol. 56, no. 6, pp. 1916-1924, Jun. 2009.

[11] Sa. Bolognani, Si. Bolognani, L. Peretti, M. Zigliotto, "Design and Implementation of Model Predictive Control for Electrical Motor Drives", *IEEE Trans. Ind. Electron.*, vol. 56, no. 6, pp. 1925-1936, Jun. 2009.

[12] K. H. Kim, M. J. Youn, "A simple and robust digital current control technique of a PM synchronous motor using time delay control approach", *IEEE Trans. Power Electron.*, vol. 16, no. 1, pp. 72-82, Jan. 2001.

[13] H. T. Moon, H. S. Kim, M. J. Youn, "A discrete-time predictive current control for PMSM", *IEEE Trans. Power Electron.*, vol. 18, no. 1, pp. 464-472, Jan. 2003.

[14] M. Pacas, J. Weber, "Predictive direct torque control for the PM synchronous machine", *IEEE Trans. Ind. Electron.*, vol. 52, no. 5, pp. 1350-1356, Oct. 2005.

[15] P. Wipasuramonton, Z.Q. Zhu, D. Howe, "Predictive current control with current-error correction for PM brushless AC drives", *IEEE Trans. Ind. Appl.*, vol. 42, no. 4, pp. 1071-1079, Jul./Aug. 2006.

[16] S. Matsutani, T. Zanma, Y. Sumiyoshi, M. Ishida, A. Imura, M. Fujitsuna, "Optimal control of PMSMs using model predictive control with integrator", *in Proc. ICROS-*

SICE International Joint Conference, Fukuoka (Japan), Aug. 18-21, 2009, pp. 4847-4852.

[17] F. Morel, X. Lin-Shi, J.-M. Retif, B. Allard, C. Buttay, "A comparative study of predictive current control schemes for a permanent-magnet synchronous machine drive", *IEEE Trans. Ind. Electron.*, vol. 56, no. 7, pp. 2715-2728, Jul. 2009.

[18] S. Matsutani, T. Zanma, Y. Sumiyoshi, M. Ishida, A. Imura, M. Fujitsuna, "Optimal control of PMSMs using model predictive control with integrator", *in Proc. ICROS-SICE International Joint Conference*, Fukuoka (Japan), Aug. 18-21, 2009, pp. 4847-4852.

[19] H. Zhu, X. Xiao, Y. Li, "Torque Ripple Reduction of the Torque Predictive Control Scheme for Permanent-Magnet Synchronous Motors", *IEEE Trans. Ind. Electron.*, vol. 59, no. 2, pp. 871-877, Feb. 2012.

[20] H. Liu, S. Li, "Speed Control for PMSM Servo System using Predictive Functional Control and Extended State Observer", *IEEE Trans. Ind. Electron.*, vol. 59, no. 2, pp. 1171-1183, Feb. 2012.

[21] E.J. Fuentes, C.A. Silva, J.I. Yuz, "Predictive Speed Control of a Two-Mass System driven by a Permanent Magnet Synchronous Motor", *IEEE Trans. Ind. Electron.*, vol. 59, no. 7, pp. 2840-2848, July 2012.

[22] R. Errouissi, M. Ouhrouche, W.H. Chen, A.M. Trzynadlowski, "Robust Nonlinear Predictive Controller for Permanent Magnet Synchronous Motors with Optimized Cost Function", *IEEE Trans. Ind. Electron.*, vol. 59, no. 7, pp. 2849-2858, July 2012.

[23] R. Errouissi, M. Ouhrouche, W.H. Chen, A.M. Trzynadlowski, "Robust Cascaded Nonlinear Predictive Control of a Permanent Magnet Synchronous Motor with Anti-windup Compensator", *IEEE Trans. Ind. Electron.*, vol. 59, no. 8, pp. 3078-3088, Aug. 2012.

[24] K. Miyashita, H. Hara, S. Shimogata, T. Yokoyama, "Multirate deadbeat control for PWM inverters using FPGA based hardware controller", *in Proc. 28th Annual Conference of the Industrial Electronics Society (IECON)*, Sevilla (Spain), Nov. 5-8, 2002, vol. 1, pp. 95-100.

[25] T. Yokoyama, M. Horiuchi, S. Shimogata, "Instantaneous deadbeat control for PWM inverter using FPGA based hardware controller", *in Proc. 29th Annual Conference of the IEEE Industrial Electronics Society (IECON),* Roanoke (USA), Nov. 2-6, 2003, vol. 1, pp.180-185.

[26] M.W. Naouar, E. Monmasson, A.A. Naassani, I. Slama-Belkhodja, N. Patin, "FPGA-based current controllers for AC machine drives – A review", *IEEE Trans. Ind. Electron.*, vol. 54, no. 4, pp. 1907-1925, Aug. 2007.

[27] A. Kawamura, H. Fujimoto, T. Yokoyama, "Survey on the real time digital feedback control of PWM inverter and the extension to multi-rate sampling and FPGA based inverter control", *in Proc. 33rd Annual Conference of the IEEE Industrial Electronics Society (IECON)*, Taipei (Taiwan), Nov. 5-8, 2007, pp. 2044-2051.

[28] E. Monmasson, L. Idkhajine, M. N. Cirstea, I. Bahri, A. Tisan, M.W. Naouar, "FPGAs in industrial control applications", *IEEE Trans. Ind. Informat.*, vol. 7, no. 2, pp. 224-243, May 2011.

[29] G. Gatto, I. Marongiu, A. Serpi, "Discrete-Time Parameter Identification of a Surface-Mounted Permanent Magnet Synchronous Machine", *IEEE Trans. Ind. Electron.,* vol. 60, no. 11, pp. 4869-4880, Nov. 2013.

[30] Sa. Bolognani, Si. Bolognani, L. Peretti, M. Zigliotto, "Combined speed and current Model Predictive Control with inherent field-weakening features for PMSM Drives", *in Proc. 14th IEEE Mediterranean Electrotechnical Conference (MELECON 2008)*, Ajaccio (France), May 5-7, 2008, pp. 472-478.

[31] M. Tursini, E. Chiricozzi, R. Petrella, "Feedforward Flux-Weakening Control of Surface-Mounted Permanent-Magnet Synchronous Motors Accounting for Resistive Voltage Drop", *IEEE Trans. Ind. Electron.*, vol. 57, no. 1, pp. 440-448, Jan. 2010.

[32] D. Lu, N. Kar, "A review of flux-weakening control in permanent magnet synchronous machines", *in Proc. IEEE Vehicle Power and Propulsion Conference (VPPC 2010)*, Lille (France), Sept. 1-3, 2010, pp. 1-6.

[33] P. Vaclavek, P. Blaha, "Field weakening in PMSM model based predictive control", *in Proc. IEEE International Conference on Power and Energy (PECon 2010)*, Kuala Lumpur (Malaysia), Nov. 29-Dec. 1, 2010, pp. 330-335.

[34] Y. Inoue, S. Morimoto, M. Sanada, "Comparative study of PMSM Drive systems based on current control and direct torque control in flux-weakening control region", *in Proc. IEEE International Electric Machines Drives Conference (IEMDC 2011)*, Niagara Falls (USA), May 15-18, 2011, pp. 1094-1099.

[35] S. Bolognani, S. Calligaro, R. Petrella, F. Pogni, "Flux-weakening in IPM motor drives: Comparison of state-of-art algorithms and a novel proposal for controller design", *in Proc. 14th European Conference on Power Electronics and Applications (EPE 2011)*, Birmingham (UK), Aug. 30-Sept. 1, 2011, pp. 1-11.

[36] H. Liu; Z. Zhu, E. Mohamed, Y. Fu, X. Qi, "Flux-Weakening Control of Nonsalient Pole PMSM Having Large Winding Inductance, Accounting for Resistive Voltage Drop and Inverter Nonlinearities", *IEEE Trans. Power Electron.*, vol. 27, no. 2, pp. 942-952, Feb. 2012.

[37] P.-Y. Lin, Y.-S. Lai, "Voltage Control Technique for the Extension of DC-Link Voltage Utilization of Finite-Speed SPMSM Drives", *IEEE Trans. Ind. Electron.*, vol. 59, no. 9, pp. 3392-3402, Sept. 2012.

[38] G. Pellegrino, E. Armando, P. Guglielmi, "Direct-Flux Vector Control of IPM Motor Drives in the Maximum Torque Per Voltage Speed Range", *IEEE Trans. Ind. Electron.*, vol. 59, no. 10, pp. 3780-3788, Oct. 2012.

[39] Y. Inoue, S. Morimoto, M. Sanada, "Control Method Suitable for Direct-Torque-Control-Based Motor Drive System Satisfying Voltage and Current Limitations", *IEEE Trans. Ind. Appl.*, vol. 48, no. 3, pp. 970-976, May/June 2012.

[40] C.-H. Choi, J.-K. Seok, R. Lorenz, "Deadbeat-direct torque and flux control for interior PM synchronous motors operating at voltage and current limits", *in Proc. IEEE Energy Conversion Congress and Exposition (ECCE 2011)*, Phoenix (USA), Sept. 17-22, 2011, pp. 371-376.

[41] C.H. Choi, J.K. Seok, R.D. Lorenz, "Wide-Speed Direct Torque and Flux Control for Interior PM Synchronous Motors Operating at Voltage and Current Limits", *IEEE Trans. Ind. Appl.*, vol. 49, no. 1, pp. 109-117, Jan./Feb. 2013.

[42] S.H. Kim, J.K. Seok, "Maximum Voltage Utilization of IPMSMs Using Modulating Voltage Scalability for Automotive Applications", *IEEE Trans. Power Electron.*, vol. 28, no. 12, pp. 5639-5646, Dec. 2013.

[43] D. Maric, S. Hiti, C. Stancu, J. Nagashima, D. Rutledge, "Two flux weakening schemes for surface-mounted permanent-magnet synchronous drives. Design and transient

response considerations", *in Proc. IEEE International Symposium on Industrial Electronics (ISIE'99)*, Bled (Slovenia), July 12-16, 1999, vol. 2, pp. 673-678.

[44] T.-S. Kwon, S.-K. Sul, "Novel Antiwindup of a Current Regulator of a Surface-Mounted Permanent-Magnet Motor for Flux-Weakening Control", *IEEE Trans. Ind. Appl.*, vol. 42, no. 5, pp. 1293-1300, Sept./Oct. 2006.

[45] Y.-C. Kwon, S. Kim, S.-K. Sul, "Voltage Feedback Current Control Scheme for Improved Transient Performance of Permanent Magnet Synchronous Machine Drives", *IEEE Trans. Ind. Electron.*, vol. 59, no. 9, pp. 3373-3382, Sept. 2012.

[46] R. Nalepa, T. Orlowska-Kowalska, "Optimum Trajectory Control of the Current Vector of a Nonsalient-Pole PMSM in the Field-Weakening Region", *IEEE Trans. Ind. Electron.*, vol. 59, no. 7, pp. 2867-2876, July 2012.

[47] C.A. Silva, J.I. Yuz, "On sampled-data models for model predictive control", *in Proc. 36th Annual Conference of the IEEE Industrial Electronics Society (IECON)*, Phoenix (AZ, USA), Nov. 7-10, 2010, pp. 2966-2971.

[48] A. Damiano, G. Gatto, I. Marongiu, A. Perfetto, A. Serpi, "Operating Constraints Management of a Surface-Mounted PM Synchronous Machine by means of an FPGA-based Model Predictive Control Algorithm", *IEEE Trans. Ind. Informat.,* vol. 10, no. 1, pp. 243-255, Feb. 2014.

[49] A. Damiano, G. Gatto, I. Marongiu, A. Perfetto, A. Serpi, "A Novel Flux-Weakening Approach for Surface-Mounted Permanent Magnet Synchronous Machines", *in Proc. 39th Annual Conference of the IEEE Industrial Electronics Society (IECON 2013),* Vienna (Austria), Nov. 10-13, 2013, pp. 2545-2550.

[50] A. Serpi, "A Flux-Weakening Predictive Control Algorithm for Extended Constant-Power Operation of Surface-Mounted PM Machines", *in Proc. XXI International Conference on Electrical Machines (ICEM 2014),* Berlin (Germany), Sept. 2-5, 2014, pp. 658-664.

Biographical Sketches

Alessandro Serpi

Name: Alessandro Serpi

Affiliation: Department of Electrical and Electronic Engineering, University of Cagliari

Date of Birth: November 23rd, 1979

Education: Master Degree in Electrical Engineering, December 9th, 2004. PhD in Industrial Engineering, February 27th, 2009

Address: Piazza d'Armi, I-09123 Cagliari, Italy

Research and Professional Experience

Alessandro Serpi is currently a research fellow at the Department of Electrical and Electronic Engineering of the University of Cagliari. His research activity mainly focuses on design and implementation of predictive control algorithms for electrical drives and power electronic converters by means of FPGA-based control boards. It also concerns the development of suitable control and/or management strategies for Electric Vehicles, especially focusing on

Vehicle-to-Grid and renewable energy sources exploitation issues. He is author of more than 40 papers published on international conference proceedings and journals. He is also involved in some research projects, among which "Development and implementation of a High-Speed Flywheel Energy Storage System for Electric Vehicles" and "Development, design and implementation of optimal management and control systems for a Micro Smart Grid".

Professional Appointments

- Professor of "Laboratory of Electrical Drives"
- Member of the IEEE and of some of its societies (IAS, IES, IMS, PELS, PES and VTS)
- Member of the editorial board of "Science Journal of Energy Engineering".
- Reviewer of some international journals and conferences, among which "IEEE Transactions on Industrial Electronics", "IEEE Transactions on Industrial Informatics", "Annual Conference of the IEEE Industrial Electronics Society" (IECON) and "International Conference on Electrical Machines" (ICEM).

Honors

- "2011 Best PhD Thesis in Power Engineering" Award, received from the IEEE PES Italian Chapter PE31, June 2013
- Second "Prize Paper Award", received from The Power Electronics Technical Committee of the IEEE Industrial Electronics Society

Publications Last 3 Years

[1] A. Damiano, I. Marongiu, M. Porru, A. Serpi, "Electric Vehicle Energy Storage Management for Renewable Energy Sources Exploitation", *in Proc. IEEE International Electric Vehicle Conference (IEVC 2012),* Greenville (USA), Mar. 4-8, 2012, 8 pp.

[2] A. Damiano, G. Gatto, I. Marongiu, A. Serpi, "A Multi-Phase PM Synchronous Generator Torque Control for Direct-Drive Wind Turbines", *in Proc. International Symposium on Power Electronics, Electrical Drives, Automation and Motion (SPEEDAM 2012),* Sorrento (Italy), June 20-22, 2012, pp. 962-968.

[3] A. Damiano, G. Gatto, I. Marongiu, A. Serpi, A. Perfetto, "A Multiphase PM Synchronous Generator Wind Turbine Control for Internal Medium Voltage DC Distribution System", *in Proc. International Symposium on Power Electronics, Electrical Drives, Automation and Motion (SPEEDAM 2012),* Sorrento (Italy), June 20-22, 2012, pp. 1256-1262.

[4] G. Brando, A. Del Pizzo, G. Gatto, I. Marongiu, A. Serpi, "Sensorless Predictive Algorithm for Permanent Magnet Brushless DC Drives", *Journal of Energy and Power Engineering (JEPE),* vol. 6, no. 7, pp. 1088-1096, July 2012.

[5] A. Damiano, G. Gatto, I. Marongiu, S. Meo, A. Perfetto, A. Serpi, "Single-Stage Grid Connected PV Inverter with Active and Reactive Power Flow Control via PSO-PR Based Current Controlled SVPWM", *International Review of Electrical Engineering (IREE),* vol. 7, no. 4, pp. 4647-4654, July-Aug. 2012.

[6] A. Damiano, G. Gatto, I. Marongiu, S. Meo, A. Perfetto, A. Serpi, "An EMI

Modelling of Heat Sinks for SMPS Applications", *International Review of Electrical Engineering (IREE),* vol. 7, no. 4, pp. 4793-4799, July-Aug. 2012.

[7] A. Damiano, G. Gatto, I. Marongiu, S. Meo, A. Perfetto, A. Serpi, "A Predictive Direct Torque Control of Induction Machines", *International Review of Electrical Engineering (IREE),* vol. 7, no. 4, pp. 4837-4844, July-Aug. 2012.

[8] A. Damiano, G. Gatto, I. Marongiu, S. Meo, A. Perfetto, A. Serpi, "A Direct-Drive Wind Turbine Control for a Wind Power Plant with an Internal DC Distribution System", *International Review of Electrical Engineering (IREE),* vol. 7, no. 4, pp. 4845-4856, July-Aug. 2012.

[9] G. Gatto, I. Marongiu, A. Perfetto, A. Serpi, I.L. Spano, "Modelling and Predictive Control of a Buck-Boost DC-DC Converter", *Journal of Energy and Power Engineering (JEPE),* vol. 6, no. 8, pp. 1276-1283, Aug. 2012.

[10] A. Damiano, G. Gatto, I. Marongiu, M. Porru, A. Serpi, "The Plug-in Electric Vehicles Role in Smart Grid Development: a Survey", *in Proc. European Electric Vehicle Congress (EEVC 2012),* Brussels (Belgium), Nov. 19-22, 2012, 12 pp.

[11] G. Gatto, I. Marongiu, A. Serpi, "Discrete-Time Parameter Identification of a Surface-Mounted Permanent Magnet Synchronous Machine", *IEEE Trans. Ind. Electron.,* vol. 60, no. 11, pp. 4869-4880, Nov. 2013.

[12] G. Gatto, I. Marongiu, A. Mocci, A. Serpi, I.L. Spano, "A Novel Continuous-Time Equivalent Circuit for Boost DC-DC Converters", *in Proc. 39th Annual Conference of the IEEE Industrial Electronics Society (IECON 2013),* Vienna (Austria), Nov. 10-13, 2013, pp. 262-267.

[13] A. Damiano, I. Marongiu, M. Porru, A. Serpi, "A Suitable PWM for DC-link Voltage Equalization of Three-Level Neutral-Point Clamped Converters", *in Proc. 39th Annual Conference of the IEEE Industrial Electronics Society (IECON 2013),* Vienna (Austria), Nov. 10-13, 2013, pp. 328-333.

[14] G. Gatto, I. Marongiu, A. Mocci, A. Serpi, I.L. Spano, "An Improved Averaged Model for Boost DC-DC Converters", *in Proc. 39th Annual Conference of the IEEE Industrial Electronics Society (IECON 2013),* Vienna (Austria), Nov. 10-13, 2013, pp. 412-417.

[15] A. Damiano, G. Gatto, I. Marongiu, A. Perfetto, A. Serpi, "A Novel Flux-Weakening Approach for Surface-Mounted Permanent Magnet Synchronous Machines", *in Proc. 39th Annual Conference of the IEEE Industrial Electronics Society (IECON 2013),* Vienna (Austria), Nov. 10-13, 2013, pp. 2547-2552.

[16] A. Damiano, G. Gatto, I. Marongiu, M. Porru, A. Serpi, "Vehicle-to-Grid Technology: State-of-the-Art and Future Scenarios", *Journal of Energy and Power Engineering (JEPE),* vol. 8, no. 1, pp. 152-165, Jan. 2014.

[17] A. Damiano, G. Gatto, I. Marongiu, A. Perfetto, A. Serpi, "Operating Constraints Management of a Surface-Mounted PM Synchronous Machine by means of an FPGA-based Model Predictive Control Algorithm", *IEEE Trans. Ind. Informat.,* vol. 10, no. 1, pp. 243-255, Feb. 2014.

[18] A. Damiano, G. Gatto, I. Marongiu, M. Porru, A. Serpi, "Real-Time Control Strategy of Energy Storage Systems for Renewable Energy Sources Exploitation", *IEEE Trans. Sustain. Energy,* vol. 5, no. 2, pp. 567-576, Apr. 2014.

[19] A. Damiano, F. Deiana, G. Fois, G. Gatto, I. Marongiu, A. Serpi, A. Perfetto, "Performance Comparison between Two-phase-on and Three-phase-on Operation of Brushless DC Drives", *in Proc. 22nd International Symposium on Power Electronics,*

Electrical Drives, Automation and Motion (SPEEDAM 2014), Ischia (Italy), June 18-20, 2014, pp. 486-491.

[20] A. Mocci, A. Serpi, I. Marongiu, G. Gatto, "A Suitable Inductor Modeling for DC-DC Converters", *in Proc. 10th Conference on Ph.D. Research in Microelectronics and Electronics (PRIME 2014),* Grenoble (France), June 30-Jul. 3, 2014, 4 pp.

[21] M. Porru, A. Serpi, I. Marongiu, A. Damiano, "An Improved DC-Link Voltage Equalization for Three-Level Neutral-Point Clamped Converters", *in Proc. 10th Conference on Ph.D. Research in Microelectronics and Electronics (PRIME 2014),* Grenoble (France), June 30-Jul. 3, 2014, 4 pp.

[22] I.L. Spano, S. Sulis, A. Serpi, I. Marongiu, G. Gatto, "EMC Characterization of Implantable Cardiac Medical Devices in an Anechoic Chamber", *in Proc. IEEE International Symposium on Electromagnetic Compatibility (EMC Europe 2014),* Gothenburg (Sweden), Sept. 1-4, 2014, pp. 872-877.

[23] I.L. Spano, A. Mocci, A. Serpi, I. Marongiu, G. Gatto, "Performance and EMC Analysis of an Interleaved PFC Boost Converter Topology", *in Proc. 49th International Universities Power Engineering Conference (UPEC 2014),* Cluj-Napoca (Romania), Sept. 2-5, 2014, 6 pp.

[24] A. Serpi, "A Flux-Weakening Predictive Control Algorithm for Extended Constant-Power Operation of Surface-Mounted PM Machines", *in Proc. XXI International Conference on Electrical Machines (ICEM 2014),* Berlin (Germany), Sept. 2-5, 2014, pp. 658-664.

[25] A. Monni, I. Marongiu, A. Serpi, A. Damiano, "Design of a Fractional Slot Multi-Phase PMSG for a Direct-Drive Wind Turbine", *in Proc. XXI International Conference on Electrical Machines (ICEM 2014),* Berlin (Germany), Sept. 2-5, 2014, pp. 2087-2093.

[26] M. Porru, A. Serpi, I. Marongiu, A. Damiano, "A Novel DC-link Voltage and Current Control Algorithm for Neutral-Point-Clamped Converters", *in Proc. 40th Annual Conference of IEEE Industrial Electronics Society (IECON 2014),* Dallas (USA), Oct. 30-Nov. 1, 2014, pp. 4521-4527.

[27] A. Mocci, A. Serpi, I. Marongiu, G. Gatto, "Enhanced Modeling of DC-DC Power Converters by means of Averaging Technique", *in Proc. 40th Annual Conference of IEEE Industrial Electronics Society (IECON 2014),* Dallas (USA), Oct. 30-Nov. 1, 2014, pp. 5101-5107.

Gianluca Gatto

Name: Gianluca Gatto
Affiliation: Department of Electrical and Electronic Engineering, University of Cagliari
Date of Birth: June 25th, 1967
Education: Master Degree in Electrical Engineering, November 15th, 1994
Address: Piazza d'Armi, IT-09123 Cagliari, Italy

Research and Professional Experience

In 1994, Gianluca Gatto received the Master degree in electrical engineering from the University of Cagliari. In the same year, he joined the Department of Electrical Engineering of University of Cagliari as Associate Researcher, contributing to the development of power electronics devices for electro hydraulics actuators supply. He worked with the INFN in projects related to the study, design and testing of electric drives for "experimental X-ray tomography" for medical and industrial applications".

From 1997 to 2014, Gianluca Gatto was an Assistant Professor of Electrical Machines, Power Electronics and Electrical Drives (ING-IND/32) at the University of Cagliari, where he is currently an Associate Professor and member of the Academic Senate.

Gianluca Gatto has authored and co-authored up to seventy technical papers. His current research interests include power electronics, electrical machines and drives, as well as electromagnetic compatibility (power quality). He has been involved in several projects in the field of power electronics devices for biomedical applications and RFId technologies for Health care applications.

Professional Appointments

University courses of Power Electronics
University courses of Electromagnetic Compatibility of Energy Systems

- Professor of "Power Electronics" and "Electromagnetic Compatibility of Energy Systems"
- Head of the Electromagnetic Compatibility laboratory of the University of Cagliari
- Member of the academic Senate
- Member of the Committee of the PhD School of Industrial Engineering
- Member of the Committee of the Master Degree on Electrical Engineering
- Member of the editorial board of "Science Journal of Energy Engineering"
- Member of the IEEE and of some of its societies (EMC, IES and PELS)
- Reviewer of the international journals: IEEE Transactions on Industrial Electronics, IEEE Transactions on Industrial Informatics and Journal of Energy and Power Engineering
- Reviewer of several international conferences

Publications Last 3 Years

[1] A. Damiano, G. Gatto, I. Marongiu, A. Serpi, "A Multi-Phase PM Synchronous Generator Torque Control for Direct-Drive Wind Turbines", *in Proc. International Symposium on Power Electronics, Electrical Drives, Automation and Motion (SPEEDAM 2012),* Sorrento (Italy), June 20-22, 2012, pp. 962-968.

[2] A. Damiano, G. Gatto, I. Marongiu, A. Serpi, A. Perfetto, "A Multiphase PM Synchronous Generator Wind Turbine Control for Internal Medium Voltage DC Distribution System", *in Proc. International Symposium on Power Electronics, Electrical Drives, Automation and Motion (SPEEDAM 2012),* Sorrento (Italy), June 20-22, 2012, pp. 1256-1262.

[3] G. Brando, A. Del Pizzo, G. Gatto, I. Marongiu, A. Serpi, "Sensorless Predictive Algorithm for Permanent Magnet Brushless DC Drives", *Journal of Energy and Power Engineering (JEPE),* vol. 6, no. 7, pp. 1088-1096, July 2012.

[4] A. Damiano, G. Gatto, I. Marongiu, S. Meo, A. Perfetto, A. Serpi, "Single-Stage Grid Connected PV Inverter with Active and Reactive Power Flow Control via PSO-PR

Based Current Controlled SVPWM", *International Review of Electrical Engineering (IREE),* vol. 7, no. 4, pp. 4647-4654, July-Aug. 2012.

[5] A. Damiano, G. Gatto, I. Marongiu, S. Meo, A. Perfetto, A. Serpi, "An EMI Modelling of Heat Sinks for SMPS Applications", *International Review of Electrical Engineering (IREE),* vol. 7, no. 4, pp. 4793-4799, July-Aug. 2012.

[6] A. Damiano, G. Gatto, I. Marongiu, S. Meo, A. Perfetto, A. Serpi, "A Predictive Direct Torque Control of Induction Machines", *International Review of Electrical Engineering (IREE),* vol. 7, no. 4, pp. 4837-4844, July-Aug. 2012.

[7] A. Damiano, G. Gatto, I. Marongiu, S. Meo, A. Perfetto, A. Serpi, "A Direct-Drive Wind Turbine Control for a Wind Power Plant with an Internal DC Distribution System", *International Review of Electrical Engineering (IREE),* vol. 7, no. 4, pp. 4845-4856, July-Aug. 2012.

[8] G. Gatto, I. Marongiu, A. Perfetto, A. Serpi, I.L. Spano, "Modelling and Predictive Control of a Buck-Boost DC-DC Converter", *Journal of Energy and Power Engineering (JEPE),* vol. 6, no. 8, pp. 1276-1283, Aug. 2012.

[9] A. Damiano, G. Gatto, I. Marongiu, M. Porru, A. Serpi, "The Plug-in Electric Vehicles Role in Smart Grid Development: a Survey", *in Proc. European Electric Vehicle Congress (EEVC 2012),* Brussels (Belgium), Nov. 19-22, 2012, 12 pp.

[10] G. Gatto, I. Marongiu, A. Serpi, "Discrete-Time Parameter Identification of a Surface-Mounted Permanent Magnet Synchronous Machine", *IEEE Trans. Ind. Electron.,* vol. 60, no. 11, pp. 4869-4880, Nov. 2013.

[11] G. Gatto, I. Marongiu, A. Mocci, A. Serpi, I.L. Spano, "A Novel Continuous-Time Equivalent Circuit for Boost DC-DC Converters", *in Proc. 39th Annual Conference of the IEEE Industrial Electronics Society (IECON 2013),* Vienna (Austria), Nov. 10-13, 2013, pp. 262-267.

[12] G. Gatto, I. Marongiu, A. Mocci, A. Serpi, I.L. Spano, "An Improved Averaged Model for Boost DC-DC Converters", *in Proc. 39th Annual Conference of the IEEE Industrial Electronics Society (IECON 2013),* Vienna (Austria), Nov. 10-13, 2013, pp. 412-417.

[13] A. Damiano, G. Gatto, I. Marongiu, A. Perfetto, A. Serpi, "A Novel Flux-Weakening Approach for Surface-Mounted Permanent Magnet Synchronous Machines", *in Proc. 39th Annual Conference of the IEEE Industrial Electronics Society (IECON 2013),* Vienna (Austria), Nov. 10-13, 2013, pp. 2547-2552.

[14] A. Damiano, G. Gatto, I. Marongiu, M. Porru, A. Serpi, "Vehicle-to-Grid Technology: State-of-the-Art and Future Scenarios", *Journal of Energy and Power Engineering (JEPE),* vol. 8, no. 1, pp. 152-165, Jan. 2014.

[15] A. Damiano, G. Gatto, I. Marongiu, A. Perfetto, A. Serpi, "Operating Constraints Management of a Surface-Mounted PM Synchronous Machine by means of an FPGA-based Model Predictive Control Algorithm", *IEEE Trans. Ind. Informat.,* vol. 10, no. 1, pp. 243-255, Feb. 2014.

[16] A. Damiano, G. Gatto, I. Marongiu, M. Porru, A. Serpi, "Real-Time Control Strategy of Energy Storage Systems for Renewable Energy Sources Exploitation", *IEEE Trans. Sustain. Energy,* vol. 5, no. 2, pp. 567-576, Apr. 2014.

[17] A. Damiano, F. Deiana, G. Fois, G. Gatto, I. Marongiu, A. Serpi, A. Perfetto, "Performance Comparison between Two-phase-on and Three-phase-on Operation of Brushless DC Drives", *in Proc. 22nd International Symposium on Power Electronics, Electrical Drives, Automation and Motion (SPEEDAM 2014),* Ischia (Italy), June 18-

20, 2014, pp. 486-491.
[18] A. Mocci, A. Serpi, I. Marongiu, G. Gatto, "A Suitable Inductor Modeling for DC-DC Converters", *in Proc. 10th Conference on Ph.D. Research in Microelectronics and Electronics (PRIME 2014),* Grenoble (France), June 30-Jul. 3, 2014, 4 pp.
[19] I.L. Spano, S. Sulis, A. Serpi, I. Marongiu, G. Gatto, "EMC Characterization of Implantable Cardiac Medical Devices in an Anechoic Chamber", *in Proc. IEEE International Symposium on Electromagnetic Compatibility (EMC Europe 2014),* Gothenburg (Sweden), Sept. 1-4, 2014, pp. 872-877.
[20] I.L. Spano, A. Mocci, A. Serpi, I. Marongiu, G. Gatto, "Performance and EMC Analysis of an Interleaved PFC Boost Converter Topology", *in Proc. 49th International Universities Power Engineering Conference (UPEC 2014),* Cluj-Napoca (Romania), Sept. 2-5, 2014, 6 pp.
[21] F.V. Caredda, A. Fanti, G. Gatto, G. Mazzarella, I.L. Spano, E. Tanzi, P.F. Orru, F. Zedda, "EM Level Evaluation in Indoor Environment", *in Proc. 49th International Universities Power Engineering Conference (UPEC 2014),* Cluj-Napoca (Romania), Sept. 2-5, 2014, 4 pp.
[22] A. Mocci, A. Serpi, I. Marongiu, G. Gatto, "Enhanced Modeling of DC-DC Power Converters by means of Averaging Technique", *in Proc. 40th Annual Conference of IEEE Industrial Electronics Society (IECON 2014),* Dallas (USA), Oct. 30-Nov. 1, 2014, pp. 5101-5107.

Alfonso Damiano

Name: Alfonso Damiano
Affiliation: Department of Electrical and Electronic Engineering, University of Cagliari
Date of Birth: November 7th, 1966
Education: Master Degree in Electrical Engineering, February 1992
Address: Piazza d'Armi, IT-09123, Cagliari, Italy

Research and Professional Experience

Alfonso Damiano joined the Department of Electrical and Electronic Engineering at the University of Cagliari as an Assistant Professor in 1994. Since 2001, he has been an Associate Professor of electrical energy management and electrical machines.

His research activity mainly focuses on electrical machines, electrical drives, RES-based energy conversion system, energy storage systems and on the development of energy management system devoted to micro and smart grid. In this framework, he has designed non-linear control algorithms for electrical drives, energy conversion and storage systems. It also developed suitable control and/or management strategies for Electric Vehicles, especially focusing on Vehicle-to-Grid and renewable energy sources exploitation issues. He is author of more than 90 papers published on international conference proceedings and journals.

Professional Appointments

- Professor of "Electrical Machines" and "Electrical Energy Management"
- Member of the IEEE and of some of its societies (IAS, IES, PES and VTS)

- Head of the Sardinia Renewable Lab – Regional Governative Research Agency
- Scientific coordinator of the Sardinian Energy Plan
- Scientific coordinator of Project Smart City A - Labelled Municipalities selected for "Sustainable Energy Europe (SEE) e Managenergy Award 2014" by the European Commission
- Reviewer of some international journals and conferences, among which "IEEE Transactions on Industrial Electronics", "IEEE Transactions on Energy Conversion", "Annual Conference of the IEEE Industrial Electronics Society" (IECON) and "International Conference on Electrical Machines" (ICEM).

Honors

- Second "Prize Paper Award" , received from The Power Electronics Technical Committee of the IEEE Industrial Electronics Society

Publications Last 3 Years

[1] A. Damiano, I. Marongiu, M. Porru, A. Serpi, "Electric Vehicle Energy Storage Management for Renewable Energy Sources Exploitation", *in Proc. IEEE International Electric Vehicle Conference (IEVC 2012),* Greenville (USA), Mar. 4-8, 2012, 8 pp.

[2] A. Damiano, G. Gatto, I. Marongiu, A. Serpi, "A Multi-Phase PM Synchronous Generator Torque Control for Direct-Drive Wind Turbines", *in Proc. International Symposium on Power Electronics, Electrical Drives, Automation and Motion (SPEEDAM 2012),* Sorrento (Italy), June 20-22, 2012, pp. 962-968.

[3] A. Damiano, G. Gatto, I. Marongiu, A. Serpi, A. Perfetto, "A Multiphase PM Synchronous Generator Wind Turbine Control for Internal Medium Voltage DC Distribution System", *in Proc. International Symposium on Power Electronics, Electrical Drives, Automation and Motion (SPEEDAM 2012),* Sorrento (Italy), June 20-22, 2012, pp. 1256-1262.

[4] A. Damiano, G. Gatto, I. Marongiu, S. Meo, A. Perfetto, A. Serpi, "Single-Stage Grid Connected PV Inverter with Active and Reactive Power Flow Control via PSO-PR Based Current Controlled SVPWM", *International Review of Electrical Engineering (IREE),* vol. 7, no. 4, pp. 4647-4654, July-Aug. 2012.

[5] A. Damiano, G. Gatto, I. Marongiu, S. Meo, A. Perfetto, A. Serpi, "An EMI Modelling of Heat Sinks for SMPS Applications", *International Review of Electrical Engineering (IREE),* vol. 7, no. 4, pp. 4793-4799, July-Aug. 2012.

[6] A. Damiano, G. Gatto, I. Marongiu, S. Meo, A. Perfetto, A. Serpi, "A Predictive Direct Torque Control of Induction Machines", *International Review of Electrical Engineering (IREE),* vol. 7, no. 4, pp. 4837-4844, July-Aug. 2012.

[7] A. Damiano, G. Gatto, I. Marongiu, S. Meo, A. Perfetto, A. Serpi, "A Direct-Drive Wind Turbine Control for a Wind Power Plant with an Internal DC Distribution System", *International Review of Electrical Engineering (IREE),* vol. 7, no. 4, pp. 4845-4856, July-Aug. 2012.

[8] A. Damiano, G. Gatto, I. Marongiu, M. Porru, A. Serpi, "The Plug-in Electric Vehicles Role in Smart Grid Development: a Survey", *in Proc. European Electric Vehicle Congress (EEVC 2012),* Brussels (Belgium), Nov. 19-22, 2012, 12 pp.

[9] A. Damiano, I. Marongiu, M. Porru, A. Serpi, "A Suitable PWM for DC-link Voltage Equalization of Three-Level Neutral-Point Clamped Converters", *in Proc. 39th Annual Conference of the IEEE Industrial Electronics Society (IECON 2013),* Vienna (Austria), Nov. 10-13, 2013, pp. 328-333.

[10] A. Damiano, G. Gatto, I. Marongiu, A. Perfetto, A. Serpi, "A Novel Flux-Weakening Approach for Surface-Mounted Permanent Magnet Synchronous Machines", *in Proc. 39th Annual Conference of the IEEE Industrial Electronics Society (IECON 2013),* Vienna (Austria), Nov. 10-13, 2013, pp. 2547-2552.

[11] A. Damiano, I. Marongiu, A. Monni, M. Porru, "Design of a 10 MW multi-phase PM synchronous generator for direct-drive wind turbines", *in Proc. 39th Annual Conference of the IEEE Industrial Electronics Society (IECON 2013),* Vienna (Austria), Nov. 10-13, 2013, pp. 5266-5270.

[12] A. Damiano, I. Marongiu, C. Musio, M. Musio, "Concentrator photovoltaic standards: Experimental analyses of technical requirements", *in Proc. 39th Annual Conference of the IEEE Industrial Electronics Society (IECON 2013),* Vienna (Austria), Nov. 10-13, 2013, pp. 8074-8079.

[13] A. Damiano, I. Marongiu, C. Musio, M. Musio, "Outdoor characterization of a Cassegrain-type concentrator photovoltaic receiver", *in Proc. 39th Annual Conference of the IEEE Industrial Electronics Society (IECON 2013),* Vienna (Austria), Nov. 10-13, 2013, pp. 8110-8115.

[14] A. Damiano, G. Gatto, I. Marongiu, M. Porru, A. Serpi, "Vehicle-to-Grid Technology: State-of-the-Art and Future Scenarios", *Journal of Energy and Power Engineering (JEPE),* vol. 8, no. 1, pp. 152-165, Jan. 2014.

[15] A. Damiano, G. Gatto, I. Marongiu, A. Perfetto, A. Serpi, "Operating Constraints Management of a Surface-Mounted PM Synchronous Machine by means of an FPGA-based Model Predictive Control Algorithm", *IEEE Trans. Ind. Informat.,* vol. 10, no. 1, pp. 243-255, Feb. 2014.

[16] A. Damiano, G. Gatto, I. Marongiu, M. Porru, A. Serpi, "Real-Time Control Strategy of Energy Storage Systems for Renewable Energy Sources Exploitation", *IEEE Trans. Sustain. Energy,* vol. 5, no. 2, pp. 567-576, Apr. 2014.

[17] A. Damiano, F. Deiana, G. Fois, G. Gatto, I. Marongiu, A. Serpi, A. Perfetto, "Performance Comparison between Two-phase-on and Three-phase-on Operation of Brushless DC Drives", *in Proc. 22nd International Symposium on Power Electronics, Electrical Drives, Automation and Motion (SPEEDAM 2014),* Ischia (Italy), June 18-20, 2014, pp. 486-491.

[18] M. Porru, A. Serpi, I. Marongiu, A. Damiano, "An Improved DC-Link Voltage Equalization for Three-Level Neutral-Point Clamped Converters", *in Proc. 10th Conference on Ph.D. Research in Microelectronics and Electronics (PRIME 2014),* Grenoble (France), June 30-Jul. 3, 2014, 4 pp.

[19] A. Monni, I. Marongiu, A. Serpi, A. Damiano, "Design of a Fractional Slot Multi-Phase PMSG for a Direct-Drive Wind Turbine", *in Proc. XXI International Conference on Electrical Machines (ICEM 2014),* Berlin (Germany), Sept. 2-5, 2014, pp. 2087-2093.

[20] D. Vincenzi, S. Baricordi, S. Calabrese, M. Musio, A. Damiano, "A Cassegrain Concentrator Photovoltaic System: comparison between dichroic and multijunction photovoltaic configurations", *in Proc. 40th Annual Conference of IEEE Industrial Electronics Society (IECON 2014),* Dallas (USA), Oct. 30-Nov. 1, 2014, pp. 1900-

1905.
[21] M. Musio, A. Damiano, "Analysis of Vehicle to Grid and Energy Storage Integration in a Virtual Power Plant", *in Proc. 40th Annual Conference of IEEE Industrial Electronics Society (IECON 2014),* Dallas (USA), Oct. 30-Nov. 1, 2014, pp. 3094-3100.
[22] M. Porru, A. Serpi, I. Marongiu, A. Damiano, "A Novel DC-link Voltage and Current Control Algorithm for Neutral-Point-Clamped Converters", *in Proc. 40th Annual Conference of IEEE Industrial Electronics Society (IECON 2014),* Dallas (USA), Oct. 30-Nov. 1, 2014, pp. 4521-4527.

Ignazio Marongiu

Name: Ignazio Marongiu
Affiliation: Department of Electrical and Electronic Engineering, University of Cagliari
Date of Birth: January 5th, 1946
Education: Master Degree in Mechanical Engineering, June 20th, 1972
Address: Piazza d'Armi, IT-09123 Cagliari, Italy

Research and Professional Experience

Ignazio Marongiu started his research activity in 1974, when he became an associate researcher in Automatic Control at the "Istituto di Elettrotecnica" of the University of Cagliari, in Italy. Subsequently, from 1976 to 1982, he taught "Electrical Machines" and "Electrical Machine Design" for electrical engineering students as an assistant professor. From 1982 to 1995, he was an Associate Professor of Electrical Machines. Since 1995, he has been Full Professor of Power Electronics, Electrical Machines and Drives at the Department of Electrical and Electronic Engineering of the University of Cagliari.

His research activity has been on theory, design, and applications of conventional and special electrical machines and drives, as well as on non-linear and adaptive control of electrical drives and power electronic converters. He is author of more than 160 papers on international journals and conference proceedings. He is also co-author of two textbooks on electrical machines. He has been scientific coordinator of several regional, national and international research projects.

Professional Appointments

- Professor of "Electrical Drives", "Electrical Actuators" and "Electric Vehicles"
- Dean of the Committee of the Master Degree Course of Energy Engineering
- Head of the Electrical Drive Laboratory at the University of Cagliari
- Member of CT301/22G Technical Committee (International Standards on Power Electronics and Electrical Drives)
- Member of the IEEE
- Member of the editorial board of "IEEE International Symposium on Power Electronics, Electrical Drives, Automation and Motion" (SPEEDAM)
- Reviewer of several international journals and conferences
- Project reviewer of MIUR (Italian Ministry of University and Research)

Honors

- Second "Prize Paper Award" , received from The Power Electronics Technical Committee of the IEEE Industrial Electronics Society

Publications Last 3 Years

[1] A. Damiano, I. Marongiu, M. Porru, A. Serpi, "Electric Vehicle Energy Storage Management for Renewable Energy Sources Exploitation", *in Proc. IEEE International Electric Vehicle Conference (IEVC 2012),* Greenville (USA), Mar. 4-8, 2012, 8 pp.

[2] A. Damiano, G. Gatto, I. Marongiu, A. Serpi, "A Multi-Phase PM Synchronous Generator Torque Control for Direct-Drive Wind Turbines", *in Proc. International Symposium on Power Electronics, Electrical Drives, Automation and Motion (SPEEDAM 2012),* Sorrento (Italy), June 20-22, 2012, pp. 962-968.

[3] A. Damiano, G. Gatto, I. Marongiu, A. Serpi, A. Perfetto, "A Multiphase PM Synchronous Generator Wind Turbine Control for Internal Medium Voltage DC Distribution System", *in Proc. International Symposium on Power Electronics, Electrical Drives, Automation and Motion (SPEEDAM 2012),* Sorrento (Italy), June 20-22, 2012, pp. 1256-1262.

[4] G. Brando, A. Del Pizzo, G. Gatto, I. Marongiu, A. Serpi, "Sensorless Predictive Algorithm for Permanent Magnet Brushless DC Drives", *Journal of Energy and Power Engineering (JEPE),* vol. 6, no. 7, pp. 1088-1096, July 2012.

[5] A. Damiano, G. Gatto, I. Marongiu, S. Meo, A. Perfetto, A. Serpi, "Single-Stage Grid Connected PV Inverter with Active and Reactive Power Flow Control via PSO-PR Based Current Controlled SVPWM", *International Review of Electrical Engineering (IREE),* vol. 7, no. 4, pp. 4647-4654, July-Aug. 2012.

[6] A. Damiano, G. Gatto, I. Marongiu, S. Meo, A. Perfetto, A. Serpi, "An EMI Modelling of Heat Sinks for SMPS Applications", *International Review of Electrical Engineering (IREE),* vol. 7, no. 4, pp. 4793-4799, July-Aug. 2012.

[7] A. Damiano, G. Gatto, I. Marongiu, S. Meo, A. Perfetto, A. Serpi, "A Predictive Direct Torque Control of Induction Machines", *International Review of Electrical Engineering (IREE),* vol. 7, no. 4, pp. 4837-4844, July-Aug. 2012.

[8] A. Damiano, G. Gatto, I. Marongiu, S. Meo, A. Perfetto, A. Serpi, "A Direct-Drive Wind Turbine Control for a Wind Power Plant with an Internal DC Distribution System", *International Review of Electrical Engineering (IREE),* vol. 7, no. 4, pp. 4845-4856, July-Aug. 2012.

[9] G. Gatto, I. Marongiu, A. Perfetto, A. Serpi, I.L. Spano, "Modelling and Predictive Control of a Buck-Boost DC-DC Converter", *Journal of Energy and Power Engineering (JEPE),* vol. 6, no. 8, pp. 1276-1283, Aug. 2012.

[10] A. Damiano, G. Gatto, I. Marongiu, M. Porru, A. Serpi, "The Plug-in Electric Vehicles Role in Smart Grid Development: a Survey", *in Proc. European Electric Vehicle Congress (EEVC 2012),* Brussels (Belgium), Nov. 19-22, 2012, 12 pp.

[11] G. Gatto, I. Marongiu, A. Serpi, "Discrete-Time Parameter Identification of a Surface-Mounted Permanent Magnet Synchronous Machine", *IEEE Trans. Ind. Electron.,* vol. 60, no. 11, pp. 4869-4880, Nov. 2013.

[12] G. Gatto, I. Marongiu, A. Mocci, A. Serpi, I.L. Spano, "A Novel Continuous-Time

Equivalent Circuit for Boost DC-DC Converters", *in Proc. 39th Annual Conference of the IEEE Industrial Electronics Society (IECON 2013),* Vienna (Austria), Nov. 10-13, 2013, pp. 262-267.

[13] A. Damiano, I. Marongiu, M. Porru, A. Serpi, "A Suitable PWM for DC-link Voltage Equalization of Three-Level Neutral-Point Clamped Converters", *in Proc. 39th Annual Conference of the IEEE Industrial Electronics Society (IECON 2013),* Vienna (Austria), Nov. 10-13, 2013, pp. 328-333.

[14] G. Gatto, I. Marongiu, A. Mocci, A. Serpi, I.L. Spano, "An Improved Averaged Model for Boost DC-DC Converters", *in Proc. 39th Annual Conference of the IEEE Industrial Electronics Society (IECON 2013),* Vienna (Austria), Nov. 10-13, 2013, pp. 412-417.

[15] A. Damiano, G. Gatto, I. Marongiu, A. Perfetto, A. Serpi, "A Novel Flux-Weakening Approach for Surface-Mounted Permanent Magnet Synchronous Machines", *in Proc. 39th Annual Conference of the IEEE Industrial Electronics Society (IECON 2013),* Vienna (Austria), Nov. 10-13, 2013, pp. 2547-2552.

[16] A. Damiano, I. Marongiu, A. Monni, M. Porru, "Design of a 10 MW multi-phase PM synchronous generator for direct-drive wind turbines", *in Proc. 39th Annual Conference of the IEEE Industrial Electronics Society (IECON 2013),* Vienna (Austria), Nov. 10-13, 2013, pp. 5266-5270.

[17] A. Damiano, I. Marongiu, C. Musio, M. Musio, "Concentrator photovoltaic standards: Experimental analyses of technical requirements", *in Proc. 39th Annual Conference of the IEEE Industrial Electronics Society (IECON 2013),* Vienna (Austria), Nov. 10-13, 2013, pp. 8074-8079.

[18] A. Damiano, I. Marongiu, C. Musio, M. Musio, "Outdoor characterization of a Cassegrain-type concentrator photovoltaic receiver", *in Proc. 39th Annual Conference of the IEEE Industrial Electronics Society (IECON 2013),* Vienna (Austria), Nov. 10-13, 2013, pp. 8110-8115.

[19] A. Damiano, G. Gatto, I. Marongiu, M. Porru, A. Serpi, "Vehicle-to-Grid Technology: State-of-the-Art and Future Scenarios", *Journal of Energy and Power Engineering (JEPE),* vol. 8, no. 1, pp. 152-165, Jan. 2014.

[20] A. Damiano, G. Gatto, I. Marongiu, A. Perfetto, A. Serpi, "Operating Constraints Management of a Surface-Mounted PM Synchronous Machine by means of an FPGA-based Model Predictive Control Algorithm", *IEEE Trans. Ind. Informat.,* vol. 10, no. 1, pp. 243-255, Feb. 2014.

[21] A. Damiano, G. Gatto, I. Marongiu, M. Porru, A. Serpi, "Real-Time Control Strategy of Energy Storage Systems for Renewable Energy Sources Exploitation", *IEEE Trans. Sustain. Energy,* vol. 5, no. 2, pp. 567-576, Apr. 2014.

[22] A. Damiano, F. Deiana, G. Fois, G. Gatto, I. Marongiu, A. Serpi, A. Perfetto, "Performance Comparison between Two-phase-on and Three-phase-on Operation of Brushless DC Drives", *in Proc. 22nd International Symposium on Power Electronics, Electrical Drives, Automation and Motion (SPEEDAM 2014),* Ischia (Italy), June 18-20, 2014, pp. 486-491.

[23] A. Mocci, A. Serpi, I. Marongiu, G. Gatto, "A Suitable Inductor Modeling for DC-DC Converters", *in Proc. 10th Conference on Ph.D. Research in Microelectronics and Electronics (PRIME 2014),* Grenoble (France), June 30-Jul. 3, 2014, 4 pp.

[24] M. Porru, A. Serpi, I. Marongiu, A. Damiano, "An Improved DC-Link Voltage Equalization for Three-Level Neutral-Point Clamped Converters", *in Proc. 10th*

Conference on Ph.D. Research in Microelectronics and Electronics (PRIME 2014), Grenoble (France), June 30-Jul. 3, 2014, 4 pp.

[25] I.L. Spano, S. Sulis, A. Serpi, I. Marongiu, G. Gatto, "EMC Characterization of Implantable Cardiac Medical Devices in an Anechoic Chamber", *in Proc. IEEE International Symposium on Electromagnetic Compatibility (EMC Europe 2014),* Gothenburg (Sweden), Sept. 1-4, 2014, pp. 872-877.

[26] I.L. Spano, A. Mocci, A. Serpi, I. Marongiu, G. Gatto, "Performance and EMC Analysis of an Interleaved PFC Boost Converter Topology", *in Proc. 49th International Universities Power Engineering Conference (UPEC 2014),* Cluj-Napoca (Romania), Sept. 2-5, 2014, 6 pp.

[27] A. Monni, I. Marongiu, A. Serpi, A. Damiano, "Design of a Fractional Slot Multi-Phase PMSG for a Direct-Drive Wind Turbine", *in Proc. XXI International Conference on Electrical Machines (ICEM 2014),* Berlin (Germany), Sept. 2-5, 2014, pp. 2087-2093.

[28] M. Porru, A. Serpi, I. Marongiu, A. Damiano, "A Novel DC-link Voltage and Current Control Algorithm for Neutral-Point-Clamped Converters", *in Proc. 40th Annual Conference of IEEE Industrial Electronics Society (IECON 2014),* Dallas (USA), Oct. 30-Nov. 1, 2014, pp. 4521-4527.

[29] A. Mocci, A. Serpi, I. Marongiu, G. Gatto, "Enhanced Modeling of DC-DC Power Converters by means of Averaging Technique", *in Proc. 40th Annual Conference of IEEE Industrial Electronics Society (IECON 2014),* Dallas (USA), Oct. 30-Nov. 1, 2014, pp. 5101-5107.

In: Model Predictive Control
Editor: Corrine Wade, pp. 41-59
ISBN: 978-1-63463-859-3
© 2015 Nova Science Publishers, Inc.

Chapter 2

A MODEL PREDICTIVE CONTROL FOR AN AGGREGATE ACTUATOR WITH A SELF-TUNING INITIAL CONDITION PROCEDURE IN COMBUSTION ENGINES

Paolo Mercorelli[1,*]***, Benedikt Haus***[1] ***and Nils Werner***[2,†]
[1]Institute of Product and Process Innovation,
Leuphana University of Lüneburg, Lüneburg, Germany
[2]Faculty of Automotive Engineering,
Ostfalia University of Applied Sciences, Wolfsburg, Germany

Abstract

This contribution deals with a hybrid actuator composed by a piezo and a hydraulic part and with a Robust Model Predictive Control (RMPC) structure combined with a feedforward control in camless engine motor applications. A combination between a feedforward control based on an inversion of the system and an MPC structure is considered. To perform a feedforward regulator an identification of the start condition of the piezo actuator is needed. This start condition of the piezo actuator is due to some structural constructive aspects which generate an offset into the piezo position. The feedforward regulator ends up being an affine function to compensate for this offset. A procedure for its identification is proposed. The idea behind the conception of the proposed new actuator is to use the advantages of both the high precision of the piezo and the force of the hydraulic part. In fact, piezoelectric actuators (PEAs) are commonly used for precision positionings, despite the fact that PEAs present nonlinearities, such as hysteresis, saturations, and creep. In the control problem such nonlinearities must be taken into account. In this paper the Preisach dynamic model with the abovementioned nonlinearities is considered together with a feedforward control combined with a RMPC. Simulations of the implementation of the MPC structure together with the feedforward regulator and the abovementioned start condition of the piezo actuator with real data are shown.

Keywords: model predictive control, actuators, combustion engines

*E-mail address: mercorelli@uni.leuphana.de (Corresponding author). Tel: +49-(0)4131-677-5571, Fax: +49-(0)4131-677-5300.
†Tel: +49-(0)5361-831615 Fax: +49-(0)5361-831602.

1. Introduction and Motivation

In the last few years, variable engine valve control in camless systems has attracted a lot of attention because of its ability to reduce pumping losses (work required to draw air into the cylinder under a part-load operation) and to increase torque performance over a wider range than the conventional spark-ignition engine. Variable valve timing also allows control of internal exhaust gas recirculation, thus improving fuel economy and reducing NOx emissions. Besides mechanical and hydraulic variable valvetrain options, electromagnetic valve actuators have been reported in the past, [1] and in [2].

The idea presented in this paper is to use a hybrid actuator composed of a piezo and a hydraulic part in order to combine the advantages of both the high precision and velocity of the piezo and the force of the hydraulic part. Moreover, piezo actuators present in general fewer problems of electromagnetic compatibility due to the quasi- absence of the inductance effects. The main advantage of PA is nanometer scale, high stiffness, and a fast response. However, since PA has a nonlinear property which is called hysteresis effect, it leads to inaccuracy in positioning control with a high precision performance. Model Predictive Control (MPC) is an optimal control approach which can effectively deal with constraints and multivariable processes in industries. Because of its advantages, MPC has been widely applied in automotive and process control communities [3]. However, the computation burden in solving the optimization problem online usually limits the MPC applications through slow dynamic systems [3]. Recently, the study of applying MPC in mechatronic systems for servo design has attracted much interest from many researchers, owing to the emerging development of microprocessor technology. Some examples have shown great promise in a broad range of mechatronic systems such as electrical motor control [4], two stage actuation system control [5], and machine tool chattering control [6]. Moreover, various advanced techniques integrating MPC for performance improvement [7] have also been developed rapidly. The main objective of this paper, in the control sense, is to show that a feedforward regulator combined with an MPC is able to control the proposed hybrid actuator in an effective, robust way, so that it can be used in an innovative valve control application. To perform a feedforward regulator an identification of the start condition of the piezo actuator is needed. This start condition of the piezo actuator is due to some structural constructive aspects which generate an offset into the piezo position. The feedforward regulator ultimately provides an affine function to compensate for this offset. The application is a well-known camless motor engine. The combination of these two techniques tries to satisfy the exigency of robustness and optimality of the whole control structure. In fact, a feedforward control structure guarantees neither the robustness nor the optimality. An MPC strategy represents a trade-off between the robustness and the optimality of the control. The paper is organized in the following way. In section 2 a description of the whole system together with some specifications of the problems with respect to the control aspects is given. Section 3 is devoted to a short mathematical description of the well-known Preisach model. Emphasis to the state domain description of the system is given. Section 4 describes the control structure which is intended to be applied. In particular, an MPC structure together with a feedforward regulator is proposed. Concerning the feedforward regulator, an emphasis to the identification of the abovementioned start condition is given. The section of simulations and conclusions closes the paper.

The Main Nomenclature

$V_{in}(t)$: input voltage
$V_z(t)$: internal piezo voltage
$i(t)$: piezo input current
R_0: input resistance in the piezo model
R_a: parasite resistance in the piezo model
C_a: parasite capacitance in the piezo model
C_z: internal capacitance in the piezo model
$x_p(t)$: internal position of the piezo part
$x(t)$: position of the piezo mass
$x_{SK}(t)$: position of the servo piston
$\mathbf{x}(k)$: discrete state vector (position, velocity of the servo piston and position, velocity of the valve piston)
$H(x_p(t), V_{in}(t))$: hysteresis characteristic of the piezo
$u_{mpc}(k)$: discrete MPC input
$\mathbf{U}_{mpc}(k)$: discrete predicted MPC input vector
$\hat{\mathbf{Y}}_h(k)$: predicted output vector
$\mathbf{Q}_p$ and $\mathbf{R}_p$: non-negative definite matrices
$M_p/3$: moving piezo mass
M_{SK}: moving servo piston mass
K_x: internal spring constant of the piezo
K: spring constant acting on the piezo
D: damping constant acting on the piezo
D_{oil}: damping constant of the oil chamber acting on the piezo
M_{SK}: mass of the servo piston
K_{SK}: spring constant acting on the servo piston
D_{SK}: damping constant acting on the servo piston
A_1: surface of the conical hydraulic transmission from the piezo side
A_2: surface of the conical hydraulic transmission ratio from the servo piston side
T_s: sampling time

2. Description of the Whole System and Some Specifications

Figure 1 shows the principle of the engine valves to be controlled. The intake valve allows air and fuel to rush into the cylinder so combustion can take place. The exhaust valve releases the spent fuel and air mixture from the cylinder. Clearly, the timing of the valve opening and closing strongly influences the engine efficiency and fuel economy. The optimal choice of the opening and closing timing depends on the simultaneous operation conditions of the engine. In conventional spark-ignition engines the valves are driven by the camshaft and their timing is fixed by the engine speed. In general, the use of electromagnetic as well as of hybrid valve actuators decouples the valve timing from the engine speed and ensures full timing variability. If, for instance, a frequency of around 6000 rpm (revolutions per minute) is considered and a distance of 10 mm must be covered, then a time interval of approximately 4 ms for the opening and closing of the valve is required. This is one of the worst practical cases in which high accelerations up to $4500m/s^2$ have to be achieved, even in the case of large disturbances due to the strong cylinder gas force acting against the exhaust valve opening. In Fig. 1, the phase diagram of the positions of the intake and exhaust valves is sketched. Two position profiles are represented: one for the intake and the other one for the exhaust valves which operate in each cylinder. It should be noted that a crossing point between the closing trajectory of the exhaust and the opening trajectory of the intake valve is present, in order to guarantee a mix between combusted gas, the new air and gasoline. Of course, the phase synchronism with the position of the piston must be also guaranteed. Figure 1 shows the new engine structure with, in evidence, a four-piezo actuator. In more general terms, this project also involves a new conception of the functionality of some parts of the motor; for instance, the throttle valve control, see [8].

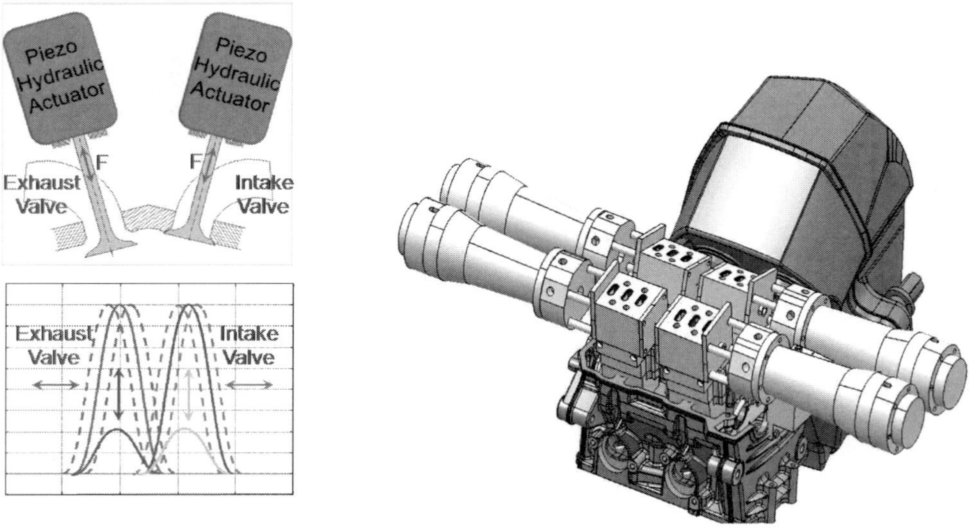

Figure 1. Exhaust and intake Piezo Hydraulic Valves and their opening and closing variable trajectories. New structure of the engine.

Figure 2 shows in detail a part of the hybrid structure which consists of a piezo actuator combined with a mechanical part. These two parts are connected by a stroke ratio to adapt the stroke length. Figure 3 shows the whole hybrid structure of the actuator. In particular, on the left side of the diagram the piezo part, which is in contact with a piston of surface equal to A_1, is visible. Through a conical structure filled with oil, the piston of surface A_1 transfers the movement to the mechanical servo piston with a surface equal to A_2. It should be noted that $A_1 >> A_2$, and this ratio allows the realization of a favorable position transfer from the piezo part to the mechanical part. Thus, the short strokes of the piezo part correspond to longer strokes of the mechanical servo part. In particular, ratio $W = \frac{A_1}{A_2} \approx 100$ for that the position is multiplied by a factor approximatively equal to 130 and as a consequence the force is reduced by the same factor. It should be noted that the piezo actuator at the elongation which is equal to 0 can produce around $30000N$ and at the elongation equal to $0.12mm$ it can produce a force of around $10000N$. In Figure 3 four hydraulic circuit points

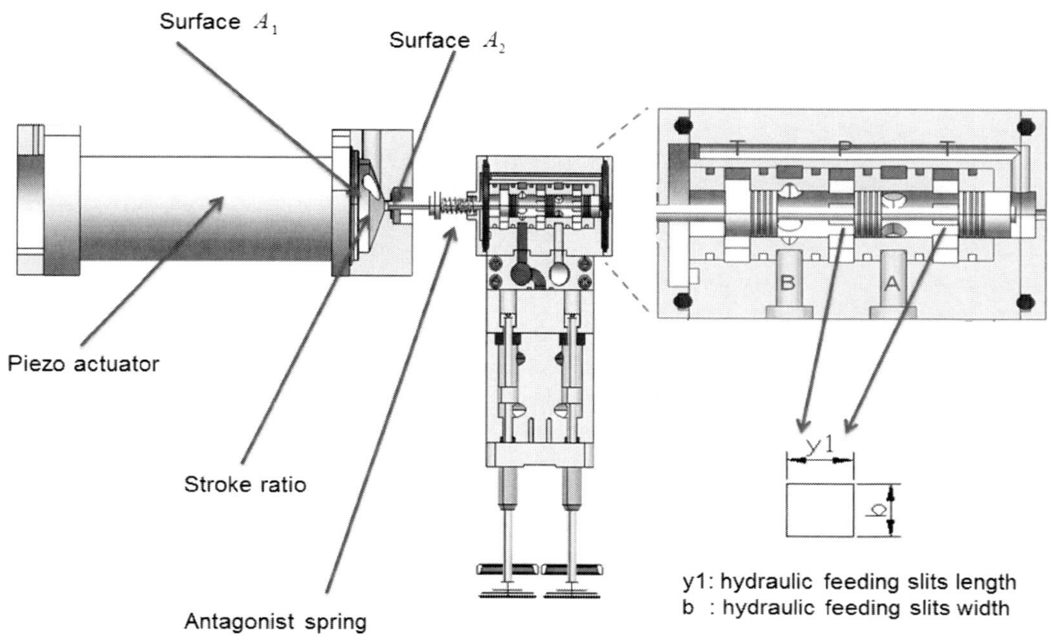

Figure 2. Structure of the hybrid actuator and details of the slits.

are visible (A-B-P-T). Through connection voids a pair of valves (either both exhaust or both intake ones) are connected to the hydraulic circuit. Once the piston 1 acts on the oil, the servo valve opens or closes the connection voids and the valves are taken into movement by the pressure exerted by the hydraulic pump through the hydraulic circuit. In more detail, in the diagram of Fig. 3 the T-A connection links the pair of valves with the tank and the P-B connection links the pair of valves with the pump. In the position of Fig. 3 connections T-A and P-B are maximally open and the pair of valves are closed because the point B is under pressure. When the piezo exerts its force, the mechanical servo valve moves and begins to close these connections. When the mechanical servo valve is in the middle position, both connections (T-A and P-B) are closed and connections A-P and B-T begin to open. At this position also both motor valves begin to open because the point A is under pressure.

3. Modeling of the Piezo Hydraulic Actuator

The proposed nonlinearity model for PEA is quite similar to those presented in [9] and in [10]. It is basically constructed from a sandwich model as shown in Fig. 4, which is based on the following hypothesis. According to the proposed sandwich model, a PEA is constructed like a three-layer sandwich. The middle layer is the effective piezo layer (P-layer), and the two outside layers connected to the electrodes are known as interfacing

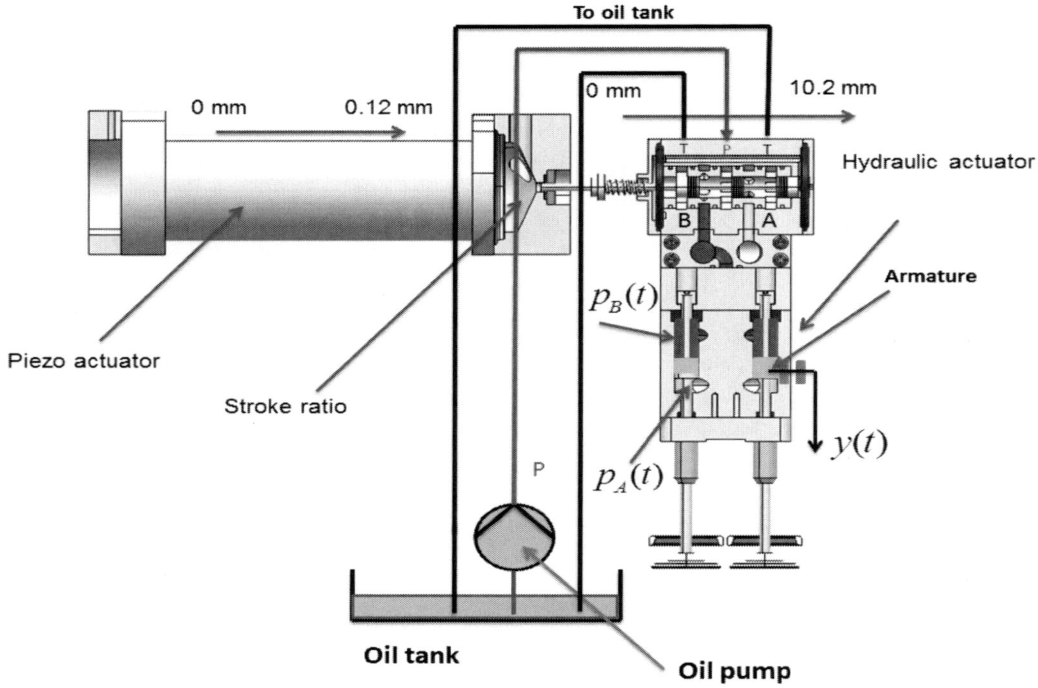

Figure 3. Scheme of the whole Hybrid Piezo Hydraulic structure.

layers (I-layers) in the literature. The P-layer is the layer that has the ordinary characteristics of piezo effects but without the nonlinearities of hysteresis and creep so that its behavior can be modeled by an equivalent linear circuitry. In contrast, the I-layers do not contribute any piezo effect; they are just parts of the circuit connecting the P-layer with the electrodes in series. In [10] it is hypothesized that each of the I-layers can be equivalently represented

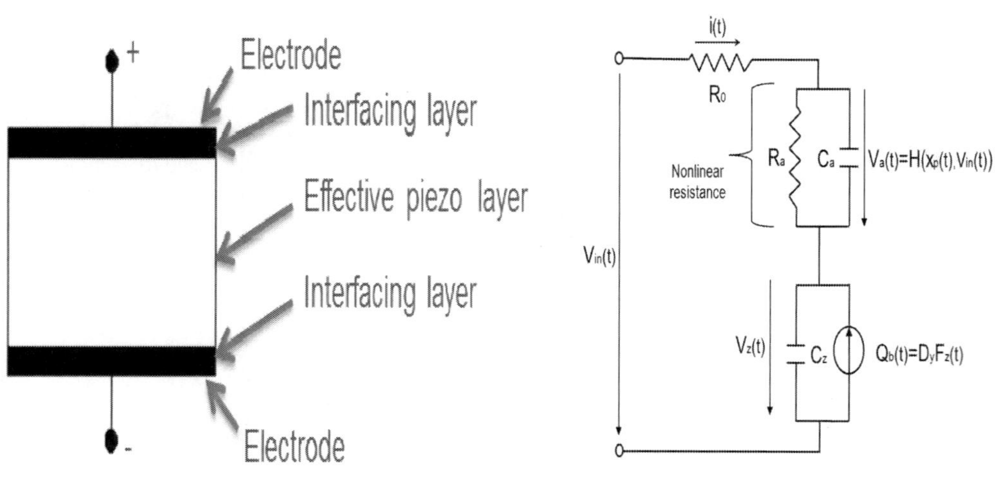

Figure 4. The sandwich model of the PEA and electrical part of the model.

by a capacitor and a resistor connected together in parallel. Together with the equivalent circuitry for P-layer, Fig. 4 shows the equivalent circuitry for a PEA with the I-layer nonlinearities of hysteresis and creep, in which two I-layers are combined together as C_a and R_a. The I-layer capacitor, C_a , is an ordinary one, which might be varied slightly with some factors, but here it would be assumed constant first for simplicity. The I-layer resistor, R_a, however, is really an extraordinary one with a significant nonlinearity. The resistance is either fairly large or fairly small. In fact, in our case the constructer gave us $R_a > 10^6\ \Omega$, when the voltage $\|V_a\| < V_h$ and $R_a < 1000$, when $\|V_a\| > V_h$. In [10], the threshold voltage, V_h, is defined as the hysteresis voltage of a PEA. Resistance R_a depends on the structure of the piezo device and on its electrical working points, voltage, current, etc. Possible values lie in a range between 50 and 1200 Ohms. This resistance changes across V_h so that it introduces the nonlinearities of hysteresis and creep in a PEA. The hysteresis effect could be seen as a function of input $V_{in}(t)$ and the motion of the piezo $x_p(t)$ as follows: $H(x_p(t), V_{in}(t))$, see Fig. 6. According to this model, if $V_h = 0$, then the hysteresis will disappear, and if $R_a = \infty$ when $\|V_a\| < V_h$, then the creep will also disappear. Based on this proposed sandwich model and the equivalent circuitry, as shown in Fig. 4, we can further derive the state model as follows:

$$\dot{V}_a(t) = -\Big(\frac{1}{R_a} + \frac{1}{R_o}\Big)\frac{V_a(t)}{C_a} - \frac{V_z(t)}{C_a R_o} + \frac{V_{in}(t)}{C_a R_o} \tag{1}$$

$$\dot{V}_z(t) = \frac{\dot{Q}_b}{C_z} + \frac{1}{C_z}\Big(-\frac{V_a(t)}{R_o} - \frac{V_z(t)}{R_o} + \frac{V_{in}(t)}{R_o}\Big), \tag{2}$$

where $Q_b = D_y F_z(t)$ is the "back electric charge force" (back-ecf) in a PEA, see [10]. According to [10] and the notation of Fig. 5, it is possible to write:

$$F_z(t) = M_p/3\ddot{x}(t) + D\dot{x}(t) + Kx(t) + K_x x(t). \tag{3}$$

K and D are the elasticity and the friction constant of the spring which is an antagonist to the piezo effect and is incorporated in the PEA. C_z is the total capacitance of the PEA and R_o is the contact resistance. For further details on this model, see [10]. Considering the whole system described in Fig. 3 with the assumptions of incompressibility of the oil, the whole mechanical system can be represented by a spring mass structure as shown in the conceptual schema of Fig. 5. In this system the following notation is adopted: K_x is the elasticity constant factor of the PEA. In the technical literature, factor $D_x K_x = T_{em}$ is known by the name "transformer ratio" and states the most important characteristic of the electromechanical transducer. $M_p/3$ is, in our case, the moving mass of the piezo structure which is a fraction of the whole piezo mass, M_{SK} is the sum of the mass of the piston with the oil and the moving actuator and M_v is the mass of the valve. It is possible to notice that the moving mass of the piezo structure is just a fraction of the whole piezo mass. The value of this fraction is given by the constructer of the piezo device and it is determined by experimental measurements. K_{SK} and D_{SK} are the characteristics of the antagonist spring to the mechanical servo valve, see Fig. 3. D_{oil} is the friction constant of the oil. Moreover, according to [10], motion $x_p(t)$ of diagram in Fig. 5 is:

$$x_p(t) = D_x V_z(t). \tag{4}$$

According to diagram of Fig. 4, it is possible to write as follows:

$$V_z = V_{in}(t) - R_0 i(t) - H(x_p(t), V_{in}(t)), \tag{5}$$

where R_0 is the connection resistance and $i(t)$ is the input current as shown in Fig. 4. $H(x_p(t), V_{in}(t))$ is the function which describes the hysteresis effect mentioned above and shown in the simulation of Fig. 6. Considering the whole system described in Fig. 3, the electrical and mechanical systems described in Figs. 4, 5 and 6 can be represented by the following mathematical expression:

$$\frac{M_p}{3}\ddot{x}(t) + M_{SK}\ddot{x}_{SK}(t) + Kx(t) + D\dot{x}(t) + K_{SK}x_{SK}(t) +$$
$$D_{SK}\dot{x}_{SK}(t) + D_{oil}\dot{x}_{SK}(t) + K_x\Big(x(t) - \Delta x_p(V_{in}(t))\Big) = 0, \tag{6}$$

where $\Delta x_p(t)$ represents the interval function of $x_p(t)$ as shown in Fig. 6 which, according to equation (4), can be expressed as:

$$\Delta x_p(t) = D_x \Delta V_z(t). \tag{7}$$

Finally, using equations (5) and (7):

$$K_x \Delta x_p(t) = K_x D_x\big(V_{in}(t) - R_0 i(t) - H(\Delta x_p(t), V_{in}(t))\big), \tag{8}$$

which represents the interval force generated by the piezo device. Equation (6) can be expressed in the following way:

$$\frac{M_p}{3}\ddot{x}(t) + M_{SK}\ddot{x}_{SK}(t) + Kx(t) + D\dot{x}(t) + K_{SK}x_{SK}(t) +$$
$$D_{SK}\dot{x}_{SK}(t) + D_{oil}\dot{x}_{SK}(t) + K_x x(t) = K_x \Delta x_p(V_{in}(t)). \tag{9}$$

It should be noted that the following relationship holds in quasi-static condition:

$$x_{SK}(t) \approx W x(t), \tag{10}$$

where W is the position ratio defined above and states the incompressibility of the oil in the conical chamber. The following equation completes the dynamic of the considered system:

$$M_v \ddot{y}(t) + D_{oil}\dot{y}(t) = F\big(x_{SK}(t), y(t)\big) - F_d(t). \tag{11}$$

According to Fig. 5, $x(t)$ is the position of the piezo actuator, $x_{SK}(t)$ is the position of the mechanical servo actuator, $y(t)$ represents the position of the valve. Function $F(x_{SK}(t), y(t))$ represents the force exerted by the pump on surface S of the armature of the moving valve, see Fig. 5. Moreover:

$$F(x_{SK}(t), y(t)) = (p_A(t) - p_B(t))S, \tag{12}$$

where $p_A(t)$ and $p_B(t)$ are the pressures in the two oil chambers separated by the armature of the valve. In [11] pressures $p_A(t)$ and $p_B(t)$ are nonlinear functions of the mechanical servo valve position $y(t)$. The linearization of this function at each desired position $y_d(t)$ will be considered later. $F_d(t)$ is the combustion back pressure in terms of force.

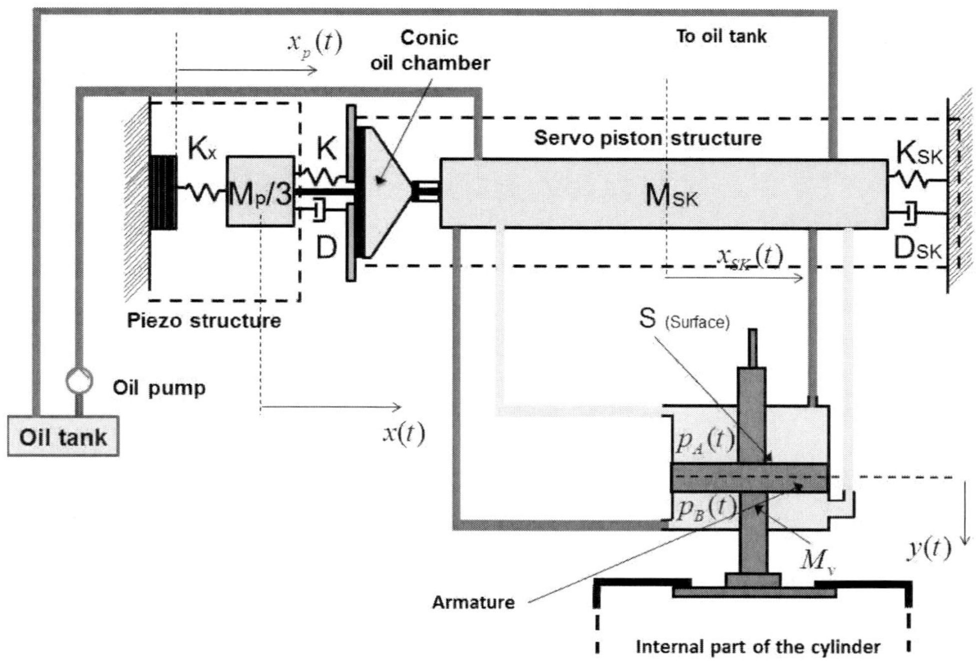

Figure 5. Graphical model of the whole actuator.

4. A Robust Representation of the System and Its Feedforward and Feedback Control

The hysteresis phenomenon could be described through a grid of affine linear functions. According to Fig. 6 an "upper and a lower bound" of this grid are clearly indicated. Through these affine linear functions of Fig. 6, it is possible to write that:

$$\Delta x_p(V_{in}(t)) = [-a \quad a] + bV_{in}(t), \tag{13}$$

with $a \in \mathbb{R}$ and $b \in \mathbb{R}$ being two positive constants which are indicated. In particular:

$$\underline{\Delta} x_p(V_{in}(t)) = -a + bV_{in}(t), \tag{14}$$

and

$$\overline{\Delta} x_p(V_{in}(t)) = a + bV_{in}(t). \tag{15}$$

Considering this notation, the system represented in (6) can be split into the following two systems:

$$\frac{M_p}{3}\ddot{x}(t) + M_{SK}\ddot{x}_{SK}(t) + Kx(t) + D\dot{x}(t) + K_{SK}x_{SK}(t) + \\ D_{SK}\dot{x}_{SK}(t) + D_{oil}\dot{x}_{SK}(t) + K_x x(t) = K_x\underline{\Delta} x_p(V_{in}(t)), \tag{16}$$

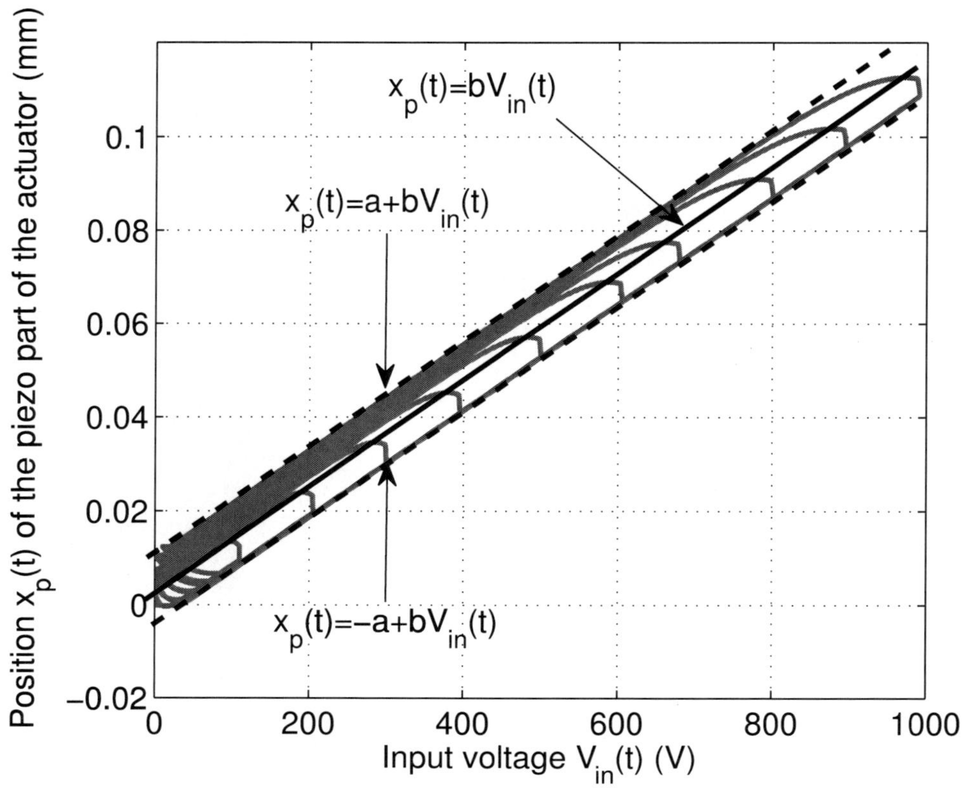

Figure 6. Simulated hysteresis curve of the piezo part of the actuator: $H(x_p(t), V_{in}(t))$.

and

$$\frac{M_p}{3}\ddot{x}(t) + M_{SK}\ddot{x}_{SK}(t) + Kx(t) + D\dot{x}(t) + K_{SK}x_{SK}(t) + D_{SK}\dot{x}_{SK}(t) + D_{oil}\dot{x}_{SK}(t) + K_x x(t) = K_x\overline{\Delta}x_p(V_{in}(t)), \quad (17)$$

where $\underline{\Delta}x_p(V_{in}(t)) = -a + bV_{in}(t)$ and $\overline{\Delta}x_p(V_{in}(t)) = a + bV_{in}(t)$.

4.1. A Self-tuning Feedforward Control Structure

Figure 7 shows the proposed control scheme and it points out a feedforward control structure which is based on the inversion of equations (16) and (17), if equation (10) is considered. In the inversion of the whole model just the mechanical part of the model is considered. In fact, according to the real data that we have, the piezo and the hydraulic part of the model are more than 10 times faster than the mechanical one. According to these considerations, which need to be further validated with simulations, the feedforward controller consists of the digitalization of the following equation:

$$u_{fed}(t) = \frac{(\frac{M_p}{3} + M_{SK}W)\ddot{y}_d(t) + \big(D + (D_{SK} + D_{oil})W\big)\dot{y}_d(t)}{K_x b} + \frac{(K + K_x + K_{SK})y_d(t) + K_x(-1)^q a}{K_x b} + u_0, \quad (18)$$

where $y_d(t)$ represents the desired trajectory position of the valve and parameter q is a switching parameter such that $q = 1,2$. In particular, it is noted that voltage u_0 represents the voltage that allows the placing of the piezo actuator in the position in which the mechanical servo valve is exactly in the middle of the oil chamber, such that the resulting pressure on the valve is equal to zero. Under these conditions the equilibrium point is reached and the valve is ready to be controlled. To clarify this and to provide more depth on this aspect, in Fig. 2 a schematic structure of a part of the system is reported. From Fig. 2 reported above, it is possible to see that when the servo mechanical piston is at the position equal to -5mm, then the pressure of the channel B equals the pressure of the tank and the pressure of the channel A equals the pressure of the hydraulic pump. When the servo mechanical piston is at the position equal to 5mm, then the pressure of the channel A equals the pressure of the tank and the pressure of the channel B equals the pressure of the hydraulic pump. At the position 0mm, the pressure in both channels equals the pressure of the tank. At this position we obtained an equilibrium of the system. This equilibrium point, which is realized by 0mm (slit length), corresponds to 5mm in the servo piston position. An MPC is proposed to make the feedforward control structure more robust.

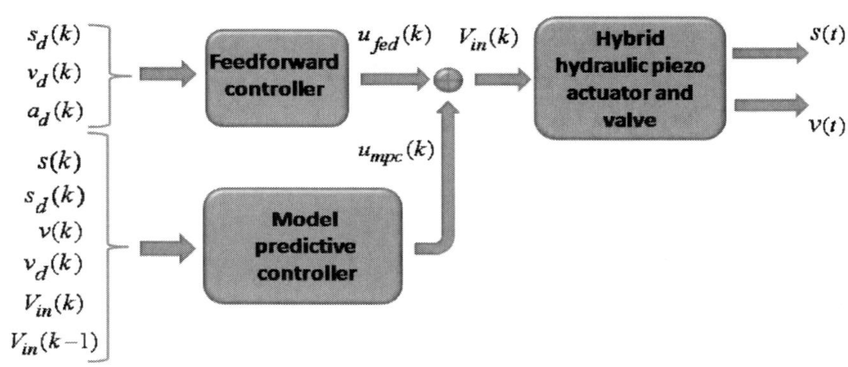

Figure 7. Whole proposed control scheme.

4.2. Solving a Linear Position MPC Optimization Problem

Considering the models described in (10), (11), (16) and (17) in which the implicit Euler discretization is considered with $k = nt_s$, $n \in \mathbb{N}$, where t_s is the sampling time, if $y(t)$ is assumed to be the controlled output, then the following system is obtained:

$$\begin{aligned} \mathbf{x}(k+1) &= \mathbf{A}_k\mathbf{x}(k) + \mathbf{B}_k V_{in}(k) + \mathbf{E}_k\mathbf{d}(k) \\ y_h(k) &= \mathbf{H}_k\mathbf{x}(k), \end{aligned} \tag{19}$$

with $\mathbf{x}(k)$ the discrete state vector consisting of the position and velocity of the servo piston together with the position and velocity of the valve piston is considered. Matrix $\mathbf{H}_k = [0 \;\; 0 \;\; 1 \;\; 0]$ is the output matrix which determines the position $y_h(k)$ of the valve according to the system in (6) and (11). Matrix $\mathbf{A}_k$, $\mathbf{B}_k$, $\mathbf{E}_k$ and $\mathbf{d}_k$ represent the digitalization of the system described in (6) and (11) according to the implicit Euler method. Particular attention should be paid to matrices $\mathbf{B}_k$, $\mathbf{E}_k$ and $\mathbf{d}_k$ which take the switching aspect in the represented system into account. For the sake of brevity, only the continuous representation of the system is shown. In fact, considering equation (11), (15) and (14) combined with (16) and (17), following **A**, **B**, **E** and **d** matrices are obtained:

$$\mathbf{A} = \begin{bmatrix} 0 & 1 & 0 & 0 \\ -\frac{K_x+K+K_{SK}W}{\frac{M_P}{3}+M_{SK}W} & -\frac{D+(D_{SK}+D_{oil})W}{\frac{M_P}{3}+M_{SK}W} & 0 & 0 \\ 0 & 0 & 0 & 1 \\ \frac{c_{x_{SK_d}(k)}}{M_v} & 0 & \frac{c_{y_d(k)}}{M_v} & -\frac{D_{oil}}{M_v} \end{bmatrix}, \tag{20}$$

$$\mathbf{B} = \begin{bmatrix} 0 \\ \frac{3K_x b}{M_p+3M_{SK}W} \\ 0 \\ 0 \end{bmatrix}, \qquad \mathbf{E} = \begin{bmatrix} 0 & 0 & 0 & 0 \\ 0 & \frac{3K_x a}{M_p+3M_{SK}W} & 0 & 0 \\ 0 & 0 & 0 & 0 \\ 0 & 0 & 0 & -\frac{1}{M_v} \end{bmatrix}, \tag{21}$$

$$\mathbf{d} = \begin{bmatrix} 0 \\ T_r(k) \\ 0 \\ -F(x_{SK_d}(k)y_d(k)) - F_L(k) \end{bmatrix}, \tag{22}$$

where $c_{x_{SK_d}(k)} \in \mathbb{R}$ and $c_{y_d(k)} \in \mathbb{R}$ are the constants which come from the linearization of $F(x_{SK}(t), y(t))$ at each k-time. $x_{SK_d}(k)$ and $y_d(k)$ are the desired trajectories of the moving mass and the desired output respectively at each k-time. The term $F(x_{SK_d}(k), y_d(k))$ comes from the linearization of $F(x_{SK}(t), y(t))$.

$$T_r = -sign(V_{in}(k) - V_{in}(k-1) - V_h(k)), \tag{23}$$

where $V_h(k)$ is a suitable threshold voltage as proposed for the construction of the hysteresis in [10]. The term T_r denotes the switching nature of the model. In the prediction model only two samples are considered:

$$\hat{y}_h(k+1/k) = \mathbf{H}_k\mathbf{A}_k\mathbf{x}(k/k) + \mathbf{H}_k\mathbf{B}_k u_{mpc}(k) + \mathbf{H}_k\mathbf{E}_k\mathbf{d}(k), \tag{24}$$

$$\hat{y}_h(k+2/k) = \mathbf{H}_k\mathbf{A}_k^2\mathbf{x}(k/k) + \mathbf{H}_k\mathbf{A}_k\mathbf{B}_k u_{mpc}(k) + \mathbf{H}_k\mathbf{B}_k u_{mpc}(k+1) + \mathbf{H}_k\mathbf{A}_k\mathbf{E}_k\mathbf{d}(k) + \mathbf{H}_k\mathbf{E}_k\mathbf{d}(k). \quad (25)$$

Equations (24) and (25) can be vectorially expressed as:

$$\hat{\mathbf{Y}}_h(k) = \mathbf{G}_p\mathbf{x}(k) + \mathbf{F}_{1p}(k)\mathbf{U}_{mpc}(k) + \mathbf{F}_{3p}\mathbf{d}(k). \quad (26)$$

where, if only two steps for the prediction horizon are considered, then:

$$\hat{\mathbf{Y}}_h(k) = \begin{bmatrix} \hat{y}_h(k+1/k) \\ \hat{y}_h(k+2/k) \end{bmatrix}, \qquad \mathbf{U}_{mpc}(k) = \begin{bmatrix} u_{mpc}(k) \\ u_{mpc}(k+1) \end{bmatrix} \quad (27)$$

and matrices $\mathbf{F}_{1p}$, $\mathbf{G}_p$ and $\mathbf{F}_{3p}$ are given here below:

$$\mathbf{F}_{1p} = \begin{bmatrix} \mathbf{H}_k\mathbf{B}_k & \mathbf{0} \\ \mathbf{H}_k(\mathbf{A}_k\mathbf{B}_k + \mathbf{B}_k) & \mathbf{H}_k\mathbf{B}_k \end{bmatrix}, \qquad \mathbf{G}_p = \begin{bmatrix} \mathbf{H}_k\mathbf{A}_k \\ \mathbf{H}_k\mathbf{A}_k^2 \end{bmatrix}, \quad (28)$$

$$\mathbf{F}_{3p} = \begin{bmatrix} \mathbf{H}_k\mathbf{E}_k & \mathbf{0} \\ \mathbf{H}_k(\mathbf{A}_k\mathbf{E}_k + \mathbf{E}_k) & \mathbf{H}_k\mathbf{E}_k \end{bmatrix}. \quad (29)$$

If the following performance criterion is assumed:

$$J = \frac{1}{2}\sum_{j=1}^{N}\Big(y_d(k+j) - \hat{y}(k+j)\Big)^{\mathrm{T}}\mathbf{Q}_p\Big(y_d(k+j)) - \hat{y}(k+j)\Big) + \frac{1}{2}\sum_{j=1}^{N-1}\Big(u_{mpc}(k+j)\Big)^{\mathrm{T}}\mathbf{R}_p u_{mpc}(k+j), \quad (30)$$

where $y_d(k+j)$, $j = 1,2,\ldots,N$ is the position reference trajectory (desired trajectory) and N the number of samples of the prediction horizon, and $\mathbf{Q}_p$ and $\mathbf{R}_p$ are non-negative definite matrices, then the solution minimizing performance index (30) may be obtained by solving:

$$\frac{\partial J}{\partial u_{mpc}} = 0. \quad (31)$$

A direct off-line computation may be obtained explicitly as:

$$u_{mpc} = (\mathbf{F}_{1p}^{\mathrm{T}}\mathbf{Q}_p\mathbf{F}_{1p} + \mathbf{R}_p)^{-1}\Big(\mathbf{F}_{1p}^{\mathrm{T}}\mathbf{Q}_p \times \big(\mathbf{Y}_{d_p}(k) - \mathbf{G}_p\mathbf{x}(k) - \mathbf{F}_{3p}\mathbf{d}(k)\big)\Big), \quad (32)$$

where $\mathbf{Y}_{d_p}(k)$ and $\mathbf{Y}_p(k)$ are the desired output column vector and the measured or observed output vector, respectively.

5. Start Condition Identification and Simulation Results

To identify the start condition of the piezo actuator the valve is opened and closed in a very short time period to cover a very short stroke amplitude. The following procedure is proposed:

Procedure

- **1.** To identify constant u_0 in equation (18) a start condition u_0^* is used, with

$$u_0^* < u_0. \tag{33}$$

- **2.** Starting with condition (33) the valve is closed and the start initial condition of the piezo actuator is incremented through a positive ramp function, see Fig. 8 with its details.
- **3.** The initial position is incremented through the positive ramp function until the velocity of the valve reaches the value equal to 0.1m/sec.
- **4.** When the velocity is equal to 0.1m/sec., then the positive ramp function is stopped and the position of the piezo actuator is saved. A negative ramp function is applied, see Fig. 8 with its details.
- **5.** Now, the initial position is decremented through the negative ramp function until the velocity of the valve reaches the value equal to -0.1m/sec.
- **6.** When the velocity is equal to -0.1m/sec., then the negative ramp function is stopped and the position of the piezo actuator is saved, see Fig. 8.
- **7.** The mean value of these two piezo actuator positions defines the start initial condition.

Figure 9 shows the final results concerning the tracking of a desired trajectory of an exhaust valve. This trajectory corresponds to 8000 rounds per minute. In Figure 10 the small movement of the piezo actuator to identify the abovementioned start position is also visible. In fact, the power involved for each exhaust valve has a peak value around 1kW and its mean value does not exceed 30 W also in case of 8000 rpm. The maximum that the current achieves, in the case of 8000 rpm, is around 1.5 Amperes. It is useful to notice that, also for a very short time, these current results are negative. This is the effect of the energy saved in the spring. This energy should be saved in a capacitor seated in the drive. Figure 11 shows the profile of the simulated reaction force on the piezo part of the proposed actuator. This force is indicated as $F_L(t)$ in Fig. 5. As already explained, because of the quasi-decoupling between the hydraulic part of the actuator and force-ratio due to the hydraulic piston, see Fig. 3, the force acting on the piezo part is strongly reduced. In fact, the force acting directly on the valve has a peak value equal to 700 N circa and it is reduced to a few Newtons acting on the piezo part due to the decoupling structure of the

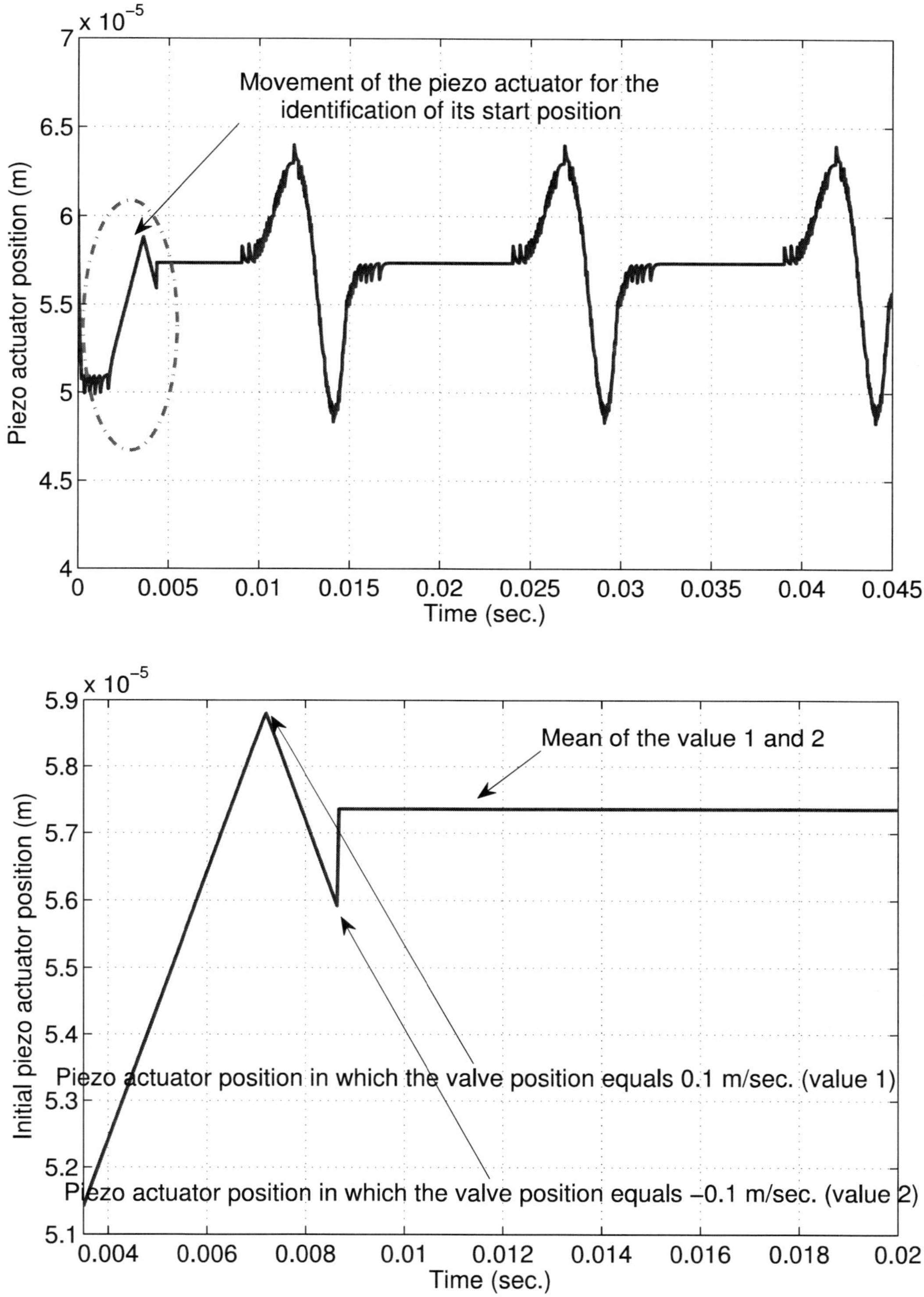

Figure 8. Piezo position with identification of its start initial condition and its detail.

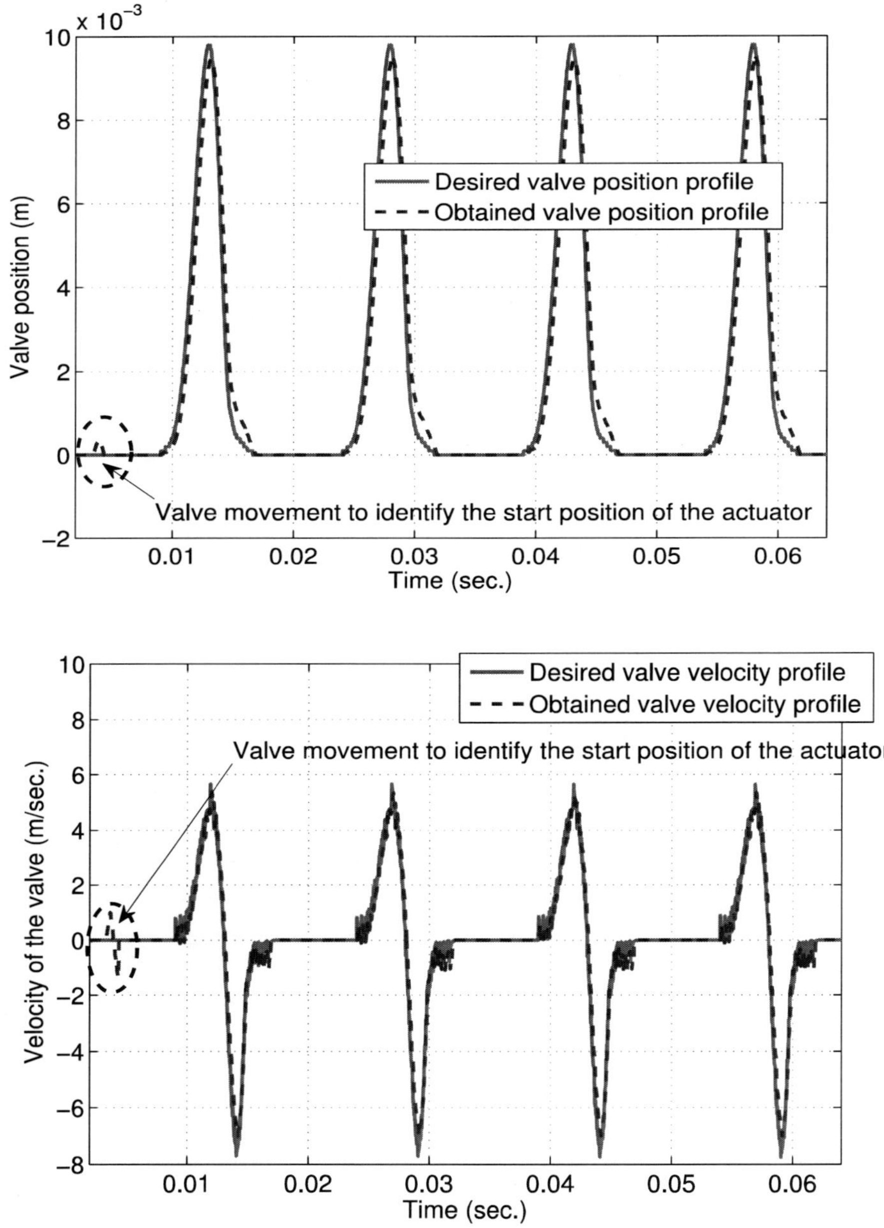

Figure 9. Position and velocity of the valve trajectory corresponding to 8000 rpm.

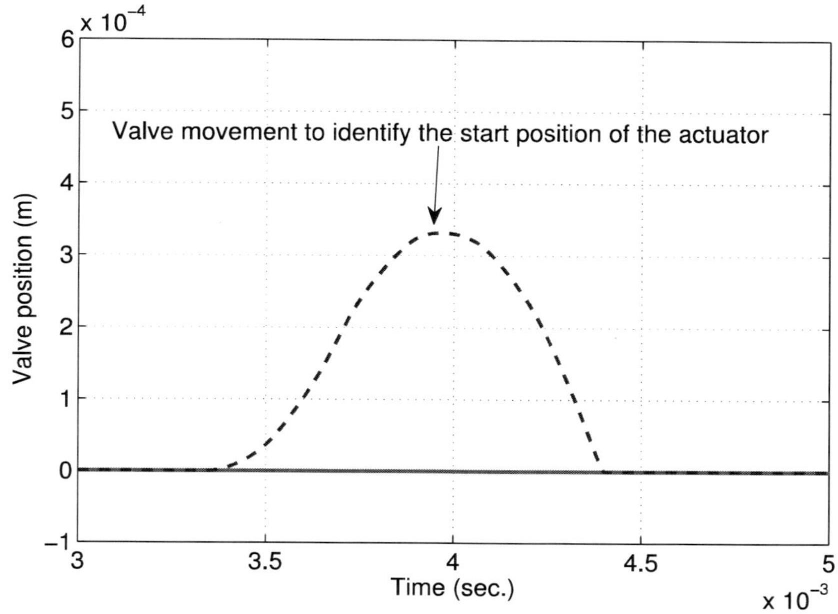

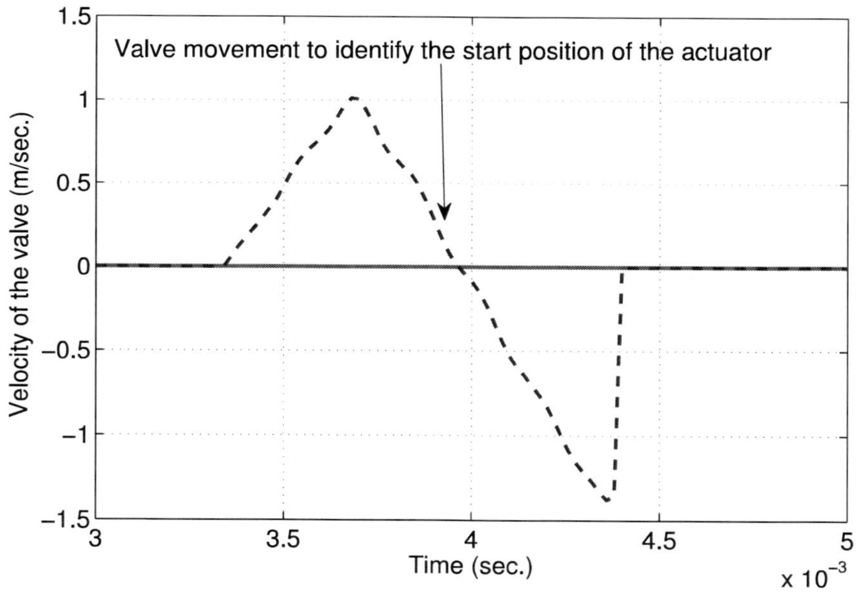

Figure 10. Details of the position and velocity of the valve trajectory corresponding to 8000 rpm.

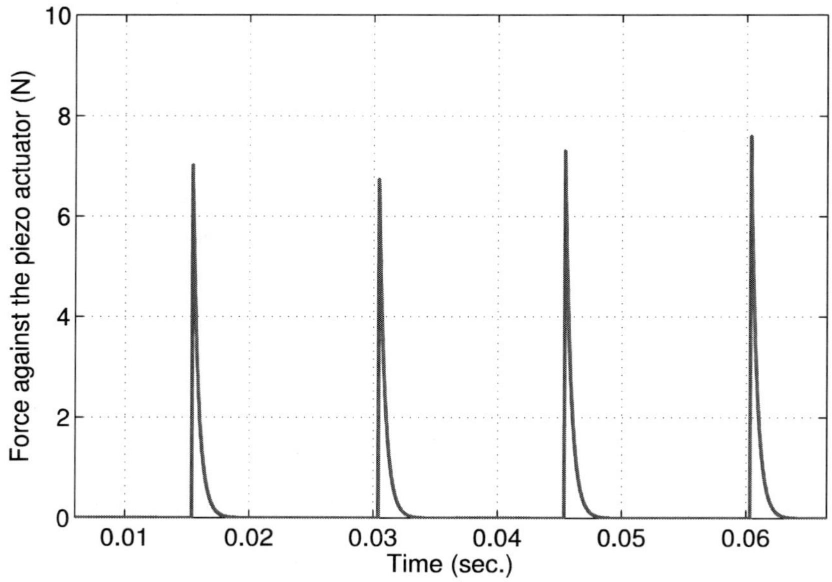

Figure 11. Force $F_L(t)$ as shown in Fig. 3: Reaction force on the internal combustion pressure corresponding to the piezo part of the actuator considering 8000 rpm.

hybrid actuator. This is one of the greatest advantages of these hybrid actuators. The model of such kind of disturbance is obtained as an exponent function of the position of the valve. The shown simulations performed a horizon consisting of 2 samples. The digital controller is set to work with a sampling time equal to $20 \times 10^{-6} sec.$ according to the specifications of the Digital Signal Processor with which we intend to test the system.

6. Conclusion and Future Objectives

6.1. Conclusion

This paper deals with a hybrid actuator composed of a piezo and a hydraulic part and its control structure for camless engine motor applications. The idea is to use the advantages of both the high precision of the piezo and the force of the hydraulic part. A combination between a feedforward control based on an inversion of the system and an MPC structure is considered. Concerning the feedforward regulator, emphasis is placed on the identification of the abovementioned start condition. In the control problem nonlinearities such as hysteresis, saturations, and creep are taken into account. Simulation results are shown.

References

[1] P. Mercorelli. A two-stage augmented extended kalman filter as an observer for sensorless valve control in camless internal combustion engines. *IEEE Transactions on Industrial Electronics*, 59(11):4236–4247, 2012.

[2] P. Mercorelli. A hysteresis hybrid extended kalman filter as an observer for sensorless valve control in camless internal combustion engines. *IEEE Transactions on Industry Applications*, 48(6):1940–1949, 2012.

[3] S.J. Qin and T.A. Badgwell. A survey of industrial model predictive control technology. *Control Engineering Practice*, 11(7):733–764, 2003.

[4] S. Bolognani, L. Peretti, and M. Zigliotto. Design and implementation of model predictive control for electrical motor drives. *IEEE Transactions on Industrial Electronics*, 56(6):1925–1936, 2009.

[5] V.A. Neelakantan, G.N. Washington, and N.K. Bucknor. Model predictive control of a two stage actuation system using piezoelectric actuators for controllable industrial and automotive brakes and clutches. *J. Intel. Mater. Syst. and Struct*, 19(7):845–857, 2008.

[6] Z. Hu and D.F. Farson. Design of a waveform tracking system for a piezoelectric actuator. In *Proc. Inst. Mech. Eng., J. Syst. Control Eng.*, volume 222, pages 11–21, 2008.

[7] R. Cao and K.S. Low. A repetitive model predictive control approach for precision tracking of a linear motion system. *IEEE Transactions on Industrial Electronics*, 56(6):1955–1962, 2009.

[8] P. Mercorelli. Robust feedback linearization using an adaptive pd regulator for a sensorless control of a throttle valve. *Mechatronics a journal of IFAC. Elsevier publishing*, 19(8):1334–1345, 2009.

[9] H.J.M.T.A. Adriaens, W. L. de Koning, and R. Banning. Modeling piezoelectric actuators. *IEEE/ASME Transactions on Mechatronics*, 5(4):331–341, 2000.

[10] Y.-C. Yu and M.-K. Lee. A dynamic nonlinearity model for a piezo-actuated positioning system. In *Proceedings of the 2005 IEEE International Conference on Mechatronics*, 2005.

[11] H. Murrenhoff. *Servohydraulik*. Shaker Verlag, Aachen, 2002.

In: Model Predictive Control
Editor: Corrine Wade, pp. 61-75
ISBN: 978-1-63463-859-3
© 2015 Nova Science Publishers, Inc.

Chapter 3

A Decoupled MPC Using a Geometric Approach and Feedforward Action for Motion Control in Robotino

Daniel Strassberger and Paolo Mercorelli*
Institute of Product and Process Innovation,
Leuphana University of Lüneburg,
Lüneburg, Germany

Abstract

Mobile robotics is a notable case of such evolution. The robotics community has developed sophisticated analysis and control techniques to meet increasing requirements for the control of motions of mechanical systems. These increasing requirements are motivated by higher performance specifications, notably an increasing number of degrees-of-freedom. This contribution proposes a controller for the motion of the Robotino. The proposed controller takes under consideration a non-interacting control strategy realized using a geometric approach. Horizontal, vertical and angular motions are considered and once the decoupling between these motions is obtained, a Model Predictive Control (MPC) strategy is used in combination with a Feedforward controller. The approach used to obtain a decoupling consists of a geometric approach. In the past three decades, research on the geometric approach to dynamic systems theory and control has allowed this approach to become a powerful and a thorough tool for the analysis and synthesis of dynamic systems. Simulation results using real data of the Robotino are shown.

Keywords: Model Predictive Control, geometric approach, decoupling

1. Introduction

In the past three decades, research on the geometric approach to dynamic systems theory and control has allowed this approach to become a powerful and thorough tool for the analysis and synthesis of dynamic systems [1], [2], [3]. Over the same time period, mechanical

*E-mail address: mercorelli@uni.leuphana.de (Corresponding author); Tel: +49-(0)4131-677-5571; Fax: +49-(0)4131-677-5300.

Figure 1. Robotino.

systems used in industry and developed in research labs have also evolved rapidly. Mobile robotics is a notable case of such evolution. The robotics community has developed sophisticated analysis and control techniques to meet increasing requirements on the control of motions of mechanical systems. These increasing requirements are motivated by higher performance specifications, specifically an increasing number of degrees-of-freedom. References [4] and [5] mark progress in the analysis and synthesis of the geometric controller for mechanical systems, and [6] proposes non-interacting force-motion control in robotic manipulation. Reference [7] reports the possibility of parameterizing input-controlled subspaces to guarantee non-interaction. In [4], a robust decoupling controller using an algebraic state input feedback is presented, while this contribution presents a robust decoupling controller using an algebraic output-input feedback. The motion control problem has attracted significant attention over the last decade in the fields of robotic manipulation and mobile robotics. Approaches exploiting decoupling controllers are found, for instance, in the work [8]. The geometric approach allows very elegant solutions to control problems. Nevertheless, robustness analysis using a linear geometric control offers answers through rank conditions of matrices that are necessary conditions. These conditions are often not constructive conditions. Even though the rank conditions offer simple "on-off" conditions, it is also possible to measure the robustness. In [4], a robust decoupling controller is obtained using a state input feedback controller. The drawback of this approach consists of a wide sensing structure. In fact, the whole state space should be available and sensed. The goal of this contribution is to propose a complete constructive procedure for the design of the decoupling controller. In the Robotino case, to achieve a decoupling between horizontal, vertical and angular motions with respect to the input forces, a preselecting field law is needed. Once the decoupling between these motions is obtained, three MPC strategies are used in combination with a preselecting matrix. Section 2 shows the model of the Robotino. Section 3 is devoted to the decoupling control design. Section 4 presents three separated MPCs for the decoupled system. Simulation results using real data close the contribution.

Figure 1 shows the Robotino in action on the pitch. This contribution is organized in the following way. Section 2 presents a possible model of the Robotino. Section 3 shows the decoupling strategy using the geometric approach. Section 4 is devoted to the derivation of the Model Predictive Control (MPC) strategy. At the end, numerical computer simulations considering real data of the Robotino are shown.

The Main Nomenclature

$\mathbf{A}$: state matrix of the mechanical model
$\mathbf{g}(\theta)$: input field of the mechanical model
$\mathbf{T}(\theta)$: decoupling input partition field
$\mathbf{T}_x(\theta)$, $\mathbf{T}_y(\theta)$, and $\mathbf{T}_\theta(\theta)$: three input partition fields
$\mathbf{B} = \mathbf{g}(\theta) \cdot \mathbf{T}(\theta)$: decoupled input matrix of the mechanical model
$\mathbf{D}(\theta)$: decoupling and stabilizing state feedback field
$\mathcal{B} = \mathsf{im}\mathbf{B}$: image of matrix $\mathbf{B}$ (subspace spanned by the columns of matrix $\mathbf{B}$)
$\mathsf{min}\mathcal{I}(\mathbf{A}, \mathcal{B}) = \sum_{i=0}^{n-1} \mathbf{A}^i \cdot \mathsf{im}\mathbf{B}$: minimum $\mathbf{A}$–invariant subspace containing $\mathsf{im}(\mathbf{B})$
$\mathcal{C} = \mathsf{ker}\mathbf{C}$: kernel of matrix $\mathbf{C}$
(subspace spanned by the columns of matrix $\mathbf{C}$)
$\mathcal{C}_x$: kernel of matrix $\mathbf{C}_x$
$\mathcal{C}_y$: kernel of matrix $\mathbf{C}_y$
$\mathcal{C}_\theta$: kernel of matrix $\mathbf{C}_\theta$
$\xi(t)$: state vector of Robotino system
$\mathbf{F}(t)$: force input vector signal
$i_1(t)$, $i_2(t)$ and $i_3(t)$: currents of the three DC-electrical drives
$\omega_1(t)$, $\omega_2(t)$ and $\omega_3(t)$: angular velocities of the three DC-electrical drives
$\mathbf{U}_{in}$: vector input voltage of the DC-drive
$U_{inp_1}(t)$, $U_{inp_2}(t)$ and $U_{inp_3}(t)$: input voltages of the DC-drive
$\mathcal{R}_x(\theta)$: minimal controlled $(\mathbf{A} + \mathbf{g}(\theta) \cdot \mathbf{D}(\theta))$-invariant subspace containing $\mathsf{im}\big(\mathbf{g}(\theta) \cdot \mathbf{T}_x(\theta)\big)$ and contained in $\mathcal{C}_y \cap \mathcal{C}_\theta$
$\mathcal{R}_y(\theta)$: minimal controlled $(\mathbf{A} + \mathbf{g}(\theta) \cdot \mathbf{D}(\theta))$-invariant subspace containing $\mathsf{im}\big(\mathbf{g}(\theta) \cdot \mathbf{T}_y(\theta)\big)$ and contained in $\mathcal{C}_x \cap \mathcal{C}_\theta$
$\mathcal{R}_\theta(\theta)$: minimal controlled $(\mathbf{A} + \mathbf{g}(\theta) \cdot \mathbf{D}(\theta))$-invariant subspace containing $\mathsf{im}\big(\mathbf{g}(\theta) \cdot \mathbf{T}_\theta(\theta)\big)$ and contained in $\mathcal{C}_x \cap \mathcal{C}_y$
$u_{mpc}(k)$: discrete MPC input
$\mathbf{U}_{mpc}(k)$: discrete predicted MPC input vector
$\hat{\mathbf{Y}}_h(k)$: predicted output vector
$\mathbf{Q}_p$ and $\mathbf{R}_p$: non-negative definite matrices
K_m: DC-motor constant
g_r: constant which takes into account the transmission ratio of the gearbox together with the axes-ratio and the radius of the wheels
L: inductance of the DC-motor
R: resistance of the DC-motor
J: motor inertia factor
K: friction factor

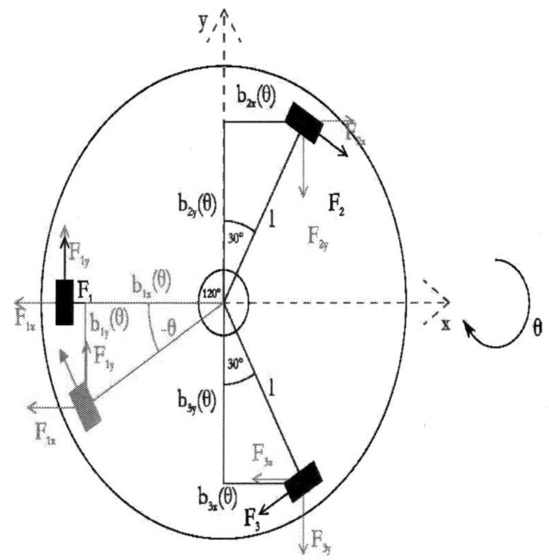

Figure 2. Mechanical schematic diagram of Robotino.

2. Mechanical and Electrical Model Description

In Fig. 2 a diagram with the representation of the forces in the considered system is shown. If state vector $\xi(t)$ is defined as $x(t)$ and $y(t)$ positions of the center of mass of the system and its velocities, $\dot{x}(t)$ and $\dot{y}(t)$, moreover considering the angular dynamics with its angular position $\theta(t)$ and its velocity $\dot{\theta}(t)$, then the following system is derived.

$$\dot{\xi}(t) = \mathbf{A}\xi(t) + \mathbf{g}(\theta)\mathbf{F}(t), \tag{1}$$

with

$$\xi(t) = \begin{bmatrix} x(t) \\ \dot{x}(t) \\ y(t) \\ \dot{y}(t) \\ \theta(t) \\ \dot{\theta}(t) \end{bmatrix}, \qquad \dot{\xi}(t) = \begin{bmatrix} \dot{\xi}(t) \\ \ddot{x}(t) \\ \dot{y}(t) \\ \ddot{y}(t) \\ \dot{\theta}(t) \\ \ddot{\theta}(t), \end{bmatrix} \tag{2}$$

matrix

$$\mathbf{A} = \begin{bmatrix} 0 & 1 & 0 & 0 & 0 & 0 \\ 0 & -\frac{k_v}{M} & 0 & 0 & 0 & 0 \\ 0 & 0 & 0 & 1 & 0 & 0 \\ 0 & 0 & 0 & -\frac{k_v}{M} & 0 & 0 \\ 0 & 0 & 0 & 0 & 0 & 1 \\ 0 & 0 & 0 & 0 & 0 & -\frac{k_\theta}{J} \end{bmatrix}, \tag{3}$$

which, for the sake of notation, can be written as follows:

$$\mathbf{A} = \begin{bmatrix} \mathbf{A}_{x2\times2} & \mathbf{0}_{2\times2} & \mathbf{0}_{2\times2} \\ \mathbf{0}_{2\times2} & \mathbf{A}_{y2\times2} & \mathbf{0}_{2\times2} \\ \mathbf{0}_{2\times2} & \mathbf{0}_{2\times2} & \mathbf{A}_{\theta2\times2} \end{bmatrix}, \tag{4}$$

field $\mathbf{g}(\theta)$ is as follows

$$\mathbf{g}(\theta) = \begin{bmatrix} 0 & 0 & 0 \\ \frac{\sin(\theta)}{M} & \frac{\sin(\frac{2}{3}\pi+\theta)}{M} & -\frac{\sin(\frac{2}{3}\pi-\theta)}{M} \\ 0 & 0 & 0 \\ \frac{\cos(\theta)}{M} & \frac{\cos(\frac{2}{3}\pi+\theta)}{M} & \frac{\cos(\frac{2}{3}\pi-\theta)}{M} \\ 0 & 0 & 0 \\ \mathbf{g}_{31}(\theta) & \mathbf{g}_{32}(\theta) & \mathbf{g}_{33}(\theta) \end{bmatrix} \tag{5}$$

where

$$\mathbf{g}_{13}(\theta) = \frac{\sin(\theta)\cdot l\cdot \sin(\theta) + \cos(\theta)\cdot l\cdot \cos(\theta)}{J}$$

$$\mathbf{g}_{23}(\theta) = \frac{\sin(\frac{2}{3}\pi+\theta)\cdot l\cdot \cos(\frac{\pi}{6}+\theta) + \cos(\frac{2}{3}\pi+\theta)\cdot l\cdot \sin(\frac{\pi}{6}+\theta)}{J}$$

$$\mathbf{g}_{33}(\theta) = \frac{\sin(\frac{2}{3}\pi-\theta)\cdot l\cdot \cos(\frac{\pi}{6}-\theta) + \cos(\frac{2}{3}\pi-\theta)\cdot l\cdot \sin(\frac{\pi}{6}-\theta)}{J}$$

and force input vector signal

$$\mathbf{F}(t) = \begin{bmatrix} F_1(t) \\ F_2(t) \\ F_3(t) \end{bmatrix}$$ in which $F_1(t) = K_m i_1(t)/g_r$, $F_2(t) = K_m i_2(t)/g_r$ and

$F_3(t) = K_m i_3(t)/g_r$, where K_m represents the DC-motor constant, g_r is a constant which takes into account the transmission ratio of the gearbox, together with the axes-ratio and the radius of the wheels. Variables $i_1(t)$, $i_2(t)$ and $i_3(t)$ are the currents of the three DC-electrical drives. In fact, the Robotino consists of three DC electrical drives that power the three omniwheels. The models of these three reported DC-drives are below.

$$\begin{bmatrix} \frac{di_1(t)}{dt} \\ \frac{d\omega_1(t)}{dt} \\ \frac{di_2(t)}{dt} \\ \frac{d\omega_2(t)}{dt} \\ \frac{di_3(t)}{dt} \\ \frac{d\omega_3(t)}{dt} \end{bmatrix} = \begin{bmatrix} -\frac{R}{L} & -\frac{K_m}{L} & 0 & 0 & 0 & 0 \\ \frac{K_m}{J} & -\frac{K}{J} & 0 & 0 & 0 & 0 \\ 0 & 0 & -\frac{R}{L} & -\frac{K_m}{L} & 0 & 0 \\ 0 & 0 & \frac{K_m}{J} & -\frac{K}{J} & 0 & 0 \\ 0 & 0 & 0 & 0 & -\frac{R}{L} & -\frac{K_m}{L} \\ 0 & 0 & 0 & 0 & \frac{K_m}{J} & -\frac{K}{J} \end{bmatrix} \cdot \begin{bmatrix} i_1(t) \\ \omega_1(t) \\ i_2(t) \\ \omega_2(t) \\ i_3(t) \\ \omega_3(t) \end{bmatrix} +$$

$$+ \begin{bmatrix} \frac{1}{L} & 0 & 0 \\ 0 & 0 & 0 \\ 0 & \frac{1}{L} & 0 \\ 0 & 0 & 0 \\ 0 & 0 & \frac{1}{L} \\ 0 & 0 & 0 \end{bmatrix} \cdot \mathbf{U}_{in}(t), \quad (6)$$

where

$$\mathbf{U}_{in}(t) = \begin{bmatrix} U_{inp_1}(t) \\ U_{inp_2}(t) \\ U_{inp_3}(t) \end{bmatrix} \quad (7)$$

with $U_{inp_1}(t)$, $U_{inp_2}(t)$ and $U_{inp_3}(t)$ represent the input voltages of the DC-drive, $\omega_1(t)$, $\omega_2(t)$ and $\omega_3(t)$ are the angular velocities of the three DC-electrical drives and L is the inductance, R is the resistance, J represents the motor inertia factor, K_m is the DC-motor constant and finally K can be seen as the friction factor.

3. Design of a Decoupling Controller

The geometric approach allows very elegant solutions to control problems. Nevertheless, robustness analysis using a linear geometric control offers answers through rank conditions of matrices that are necessary conditions. These conditions are often not constructive ones, see [1]. This section describes the design of a decoupling controller with respect to the above-defined x-y and θ motions. A geometric approach is used in this analysis. The earliest geometric approaches to non-interacting control were due to Basile and Marro ([9], [1]) and to Wonham and Morse ([10], [11], and [3]).

Definition 1. *A control law for the dynamic system (1) is* decoupling *with respect to the regulated outputs $x(t)$, $y(t)$, and $\theta(t)$, if there exist partitions $F_1(t)$, $F_2(t)$, and $F_3(t)$ of the input vector $\mathbf{F}(t)$ such that for zero initial conditions, each input $F_{(\cdot)}(t)$ (with all other inputs, identically zero) only affects the corresponding output $F_{(\cdot)}(t)$.* □

It needs to be shown that there exists a decoupling and stabilizing state feedback field $\mathbf{D}(\theta)$, along with three input partition fields $\mathbf{T}_x(\theta)$, $\mathbf{T}_y(\theta)$, and $\mathbf{T}_\theta(\theta)$ such that, for the dynamic triples

$$\begin{aligned}&\big(\mathbf{C}_x,\ \mathbf{A}+\mathbf{g}(\theta)\cdot\mathbf{D}(\theta),\ \mathbf{g}(\theta)\cdot\mathbf{F}_1(t)\big),\\&\big(\mathbf{C}_y,\ \mathbf{A}+\mathbf{g}(\theta)\cdot\mathbf{D}(\theta),\ \mathbf{g}(\theta)\cdot\mathbf{F}_2(t)\big),\\&\big(\mathbf{C}_\theta,\ \mathbf{A}+\mathbf{g}(\theta)\cdot\mathbf{D}(\theta),\ \mathbf{g}(\theta)\cdot\mathbf{F}_3(t)\big),\end{aligned} \tag{8}$$

the following conditions hold:

$$\mathcal{R}_x(\theta)=\mathsf{min}\mathcal{I}\Big(\mathbf{A}+\mathbf{g}(\theta)\cdot\mathbf{D}(\theta),\ \mathsf{im}\big(\mathbf{g}(\theta)\cdot\mathbf{T}_x(\theta)\big)\Big)\subseteq\mathcal{C}_y\cap\mathcal{C}_\theta\ \ \forall\theta, \tag{9}$$

and

$$\mathbf{C}_x\cdot\mathcal{R}_x(\theta)=\mathsf{im}(\mathbf{C}_x),\ \ \forall\theta. \tag{10}$$

$$\mathcal{R}_y(\theta)=\mathsf{min}\mathcal{I}\Big(\mathbf{A}+\mathbf{g}(\theta)\cdot\mathbf{D}(\theta),\ \mathsf{im}\big(\mathbf{g}(\theta)\cdot\mathbf{T}_y(\theta)\big)\Big)\subseteq\mathcal{C}_x\cap\mathcal{C}_\theta\ \ \forall\theta, \tag{11}$$

and

$$\mathbf{C}_y\cdot\mathcal{R}_y(\theta)=\mathsf{im}(\mathbf{C}_y),\ \ \forall\theta. \tag{12}$$

$$\mathcal{R}_\theta(\theta)=\mathsf{min}\mathcal{I}\Big(\mathbf{A}+\mathbf{g}(\theta)\cdot\mathbf{D}(\theta),\ \mathsf{im}\big(\mathbf{g}(\theta)\cdot\mathbf{T}_\theta(\theta)\big)\Big)\subseteq\mathcal{C}_x\cap\mathcal{C}_y\ \ \forall\theta, \tag{13}$$

and

$$\mathbf{C}_\theta\cdot\mathcal{R}_\theta(\theta)=\mathsf{im}(\mathbf{C}_\theta),\ \ \forall\theta. \tag{14}$$

Here,

$$\mathsf{min}\mathcal{I}(\mathbf{A},\mathsf{im}(\mathbf{g}(\theta)))=\sum_{i=0}^{n-1}\mathbf{A}^i\cdot\mathsf{im}(\mathbf{g}(\theta))$$

is a minimum $\mathbf{A}$–invariant subspace containing $\mathsf{im}(\mathbf{g}(\theta))\ \forall\ \theta$. Moreover, the partition fields $\mathbf{T}_x(\theta)$, $\mathbf{T}_y(\theta)$ and $\mathbf{T}_\theta(\theta)$ satisfy the following relationships:

$$\begin{aligned}&\mathsf{im}(\mathbf{g}(\theta)\cdot\mathbf{T}_x(\theta))=\mathsf{im}(\mathbf{g}(\theta))\cap\mathcal{R}_x(\theta),\\&\mathsf{im}(\mathbf{g}(\theta)\cdot\mathbf{T}_y(\theta))=\mathsf{im}(\mathbf{g}(\theta))\cap\mathcal{R}_y(\theta),\\&\mathsf{im}(\mathbf{g}(\theta)\cdot\mathbf{T}_\theta(\theta))=\mathsf{im}(\mathbf{g}(\theta))\cap\mathcal{R}_\theta(\theta).\end{aligned} \tag{15}$$

The stabilizing field $\mathbf{D}(\theta)$ is such that:

$$(\mathbf{A}+\mathbf{g}(\theta)\cdot\mathbf{D}(\theta))\cdot\mathcal{R}_x(\theta)\subseteq\mathcal{R}_x(\theta), \tag{16}$$

$$(\mathbf{A}+\mathbf{g}(\theta)\cdot\mathbf{D}(\theta))\cdot\mathcal{R}_y(\theta)\subseteq\mathcal{R}_y(\theta), \tag{17}$$

and

$$(\mathbf{A}+\mathbf{g}(\theta)\cdot\mathbf{D}(\theta))\cdot\mathcal{R}_\theta(\theta)\subseteq\mathcal{R}_\theta(\theta). \tag{18}$$

Considering matrix $\mathbf{A}$ defined in (3), it is straightforward to see that this matrix is structurally already decoupled and intrinsically stable; this implies that field $\mathbf{D}(\theta)=\mathbf{0}\ \forall\theta$. Considering

$$\mathbf{T}(\theta)=[\mathbf{T}_x(\theta),\mathbf{T}_y(\theta),\mathbf{T}_\theta(\theta),\mathbf{T}_c(\theta)],$$

where $\mathbf{T}_c(\theta)$ is defined in a complementary fashion, it is straightforward to show that matrix $\mathbf{T}_c = \mathbf{0}$, then

$$\mathsf{im}\mathbf{T}(\theta) = \mathsf{im}[\mathbf{T}_x(\theta), \mathbf{T}_y(\theta), \mathbf{T}_\theta(\theta)] = \mathsf{im}\mathbf{T}_x(\theta) \oplus \mathsf{im}\mathbf{T}_y(\theta) \oplus \mathsf{im}\mathbf{T}_\theta(\theta). \quad (19)$$

Considering the output matrices, which correspond to the x-position, y-position and angular position as follows:

$$\mathbf{C}_x = \begin{bmatrix} 1 & 0 & 0 & 0 & 0 & 0 \end{bmatrix}, \qquad \mathbf{C}_y = \begin{bmatrix} 0 & 0 & 1 & 0 & 0 & 0 \end{bmatrix}, \quad (20)$$

$$\mathbf{C}_\theta = \begin{bmatrix} 0 & 0 & 0 & 0 & 1 & 0 \end{bmatrix}, \quad (21)$$

which have the following kernels respectively:

$$\mathcal{C}_x = \mathsf{im} \begin{bmatrix} 0 & 0 & 0 & 0 & 0 \\ -1 & 0 & 0 & 0 & 0 \\ 0 & -1 & 0 & 0 & 0 \\ 0 & 0 & -1 & 0 & 0 \\ 0 & 0 & 0 & -1 & 0 \\ 0 & 0 & 0 & 0 & -1 \end{bmatrix}, \qquad \mathcal{C}_y = \mathsf{im} \begin{bmatrix} 1 & 0 & 0 & 0 & 0 \\ 0 & -1 & 0 & 0 & 0 \\ 0 & 0 & 0 & 0 & 0 \\ 0 & 0 & -1 & 0 & 0 \\ 0 & 0 & 0 & -1 & 0 \\ 0 & 0 & 0 & 0 & 1 \end{bmatrix}, \quad (22)$$

$$\mathcal{C}_\theta = \mathsf{im} \begin{bmatrix} 1 & 0 & 0 & 0 & 0 \\ 0 & 1 & 0 & 0 & 0 \\ 0 & 0 & 1 & 0 & 0 \\ 0 & 0 & 0 & -1 & 0 \\ 0 & 0 & 0 & 0 & 0 \\ 0 & 0 & 0 & 0 & 1 \end{bmatrix}. \quad (23)$$

According to relation (5) it is straightforward to observe that the following three equations hold $\forall\ \theta$;

$$\mathsf{im}(\mathbf{g}(\theta)) \cap \left(\mathcal{C}_x \cap \mathcal{C}_y\right) \neq \mathbf{0}, \quad (24)$$

$$\mathsf{im}(\mathbf{g}(\theta)) \cap \left(\mathcal{C}_x \cap \mathcal{C}_\theta\right) \neq \mathbf{0}, \quad (25)$$

$$\mathsf{im}(\mathbf{g}(\theta)) \cap \left(\mathcal{C}_y\right) \cap \mathcal{C}_\theta\big) \neq \mathbf{0}. \quad (26)$$

Considering that

$$(\mathbf{g}(\theta))^\dagger = \begin{bmatrix} 0 & 0 & 0 & a(\theta) & 0 & b(\theta) \\ 0 & c(\theta) & 0 & -a(\theta) & 0 & b(\theta) \\ 0 & -c(\theta) & 0 & -a(\theta) & 0 & b(\theta) \end{bmatrix}, \quad (27)$$

where with $(\mathbf{g}(\theta))^\dagger$ the pseudo inverse of field $\mathbf{g}(\theta))$ and functions $a(\theta)$, $b(\theta)$ and $c(\theta)$ are functions of the variable θ, the following calculation is obtained for $\theta \neq \pi/2 + k\pi$ with $k \in \mathbb{N}$:

$$\mathcal{C}_x \cap \mathcal{C}_y = \mathsf{im} \begin{bmatrix} 0 & 0 & 0 \\ 1 & 0 & 0 \\ 0 & 0 & 0 \\ 0 & -1 & 0 \\ 0 & 0 & 0 \\ 0 & 0 & 1 \end{bmatrix}, \mathcal{C}_\theta \cap \mathcal{C}_y \quad = \mathsf{im} \begin{bmatrix} 0 & 0 & 0 \\ 1 & 0 & 0 \\ 0 & 0 & 0 \\ 0 & 1 & 0 \\ 0 & 0 & 0 \\ 0 & 0 & 1 \end{bmatrix}, \mathcal{C}_\theta \cap \mathcal{C}x = \mathsf{im} \begin{bmatrix} 0 & 0 & 0 \\ 1 & 0 & 0 \\ 0 & 0 & 0 \\ 0 & 1 & 0 \\ 0 & 0 & 0 \\ 0 & 0 & 1 \end{bmatrix}. \quad (28)$$

The following calculation allows us to obtain the required fields for the decoupling of the mechanical system.

$$\mathbf{T}_\theta(\theta) = (\mathbf{g}(\theta))^\dagger \cdot \mathsf{im}(\mathbf{g}(\theta)) \cap \mathcal{C}_x \cap \mathcal{C}_y =$$

$$\begin{bmatrix} 0 & 0 & 0 & a(\theta) & 0 & b(\theta) \\ 0 & c(\theta) & 0 & -a(\theta) & 0 & b(\theta) \\ 0 & -c(\theta) & 0 & -a(\theta) & 0 & b(\theta) \end{bmatrix} \cdot \begin{bmatrix} 0 & 0 & 0 \\ 1 & 0 & 0 \\ 0 & 0 & 0 \\ 0 & -1 & 0 \\ 0 & 0 & 0 \\ 0 & 0 & 1 \end{bmatrix} = \begin{bmatrix} 0 & -a(\theta) & b(\theta) \\ c(\theta) & a(\theta) & b(\theta) \\ -c(\theta) & a(\theta) & b(\theta) \end{bmatrix}, \tag{29}$$

$$\mathbf{T}_x(\theta) = (\mathbf{g}(\theta))^\dagger \cdot \mathsf{im}(\mathbf{g}(\theta)) \cap \mathcal{C}_\theta \cap \mathcal{C}_y =$$

$$\begin{bmatrix} 0 & 0 & 0 & a(\theta) & 0 & b(\theta) \\ 0 & c(\theta) & 0 & -a(\theta) & 0 & b(\theta) \\ 0 & -c(\theta) & 0 & -a(\theta) & 0 & b(\theta) \end{bmatrix} \cdot \begin{bmatrix} 0 & 0 & 0 \\ 1 & 0 & 0 \\ 0 & 0 & 0 \\ 0 & 1 & 0 \\ 0 & 0 & 0 \\ 0 & 0 & 1 \end{bmatrix} = \begin{bmatrix} 0 & a(\theta) & b(\theta) \\ c(\theta) & -a(\theta) & b(\theta) \\ -c(\theta) & -a(\theta) & b(\theta) \end{bmatrix}, \tag{30}$$

$$\mathbf{T}_y(\theta) = (\mathbf{g}(\theta))^\dagger \cdot \mathsf{im}(\mathbf{g}(\theta)) \cap \mathcal{C}_\theta) \cap \mathcal{C}_x =$$

$$\begin{bmatrix} 0 & 0 & 0 & a(\theta) & 0 & b(\theta) \\ 0 & c(\theta) & 0 & -a(\theta) & 0 & b(\theta) \\ 0 & -c(\theta) & 0 & -a(\theta) & 0 & b(\theta) \end{bmatrix} \cdot \begin{bmatrix} 0 & 0 & 0 \\ 1 & 0 & 0 \\ 0 & 0 & 0 \\ 0 & 1 & 0 \\ 0 & 0 & 0 \\ 0 & 0 & 1 \end{bmatrix} = \begin{bmatrix} 0 & a(\theta) & b(\theta) \\ c(\theta) & -a(\theta) & b(\theta) \\ -c(\theta) & -a(\theta) & b(\theta) \end{bmatrix}. \tag{31}$$

Adding all three T-Fields together, we get a new field $\mathbf{T}(\theta)$:

$$\mathbf{T}(\theta) = \mathbf{T}_x(\theta) + \mathbf{T}_y(\theta) + \mathbf{T}_\theta(\theta) = \begin{bmatrix} 0 & a(\theta) & 3b(\theta) \\ 3c(\theta) & -a(\theta) & 3b(\theta) \\ -3c(\theta) & -a(\theta) & 3b(\theta) \end{bmatrix}. \tag{32}$$

Field $\mathbf{T}(\theta)$ can be seen as a preselecting field and the following product realizes the mechanical decoupling:

$$\mathcal{B} = \mathsf{im}(\mathbf{g}(\theta)\mathbf{T}(\theta)) = \mathcal{C}_x \cap \mathcal{C}_y = \mathcal{C}_\theta \cap \mathcal{C}_y = \mathcal{C}_\theta \cap \mathcal{C}_x = \mathsf{im} \begin{bmatrix} 0 & 0 & 0 \\ 1 & 0 & 0 \\ 0 & 0 & 0 \\ 0 & 1 & 0 \\ 0 & 0 & 0 \\ 0 & 0 & 1 \end{bmatrix}, \tag{33}$$

in which matrix $\mathbf{B}$ can be seen as a resulting input matrix.

4. Solving a Centralized Decoupled Linear Position MPC Optimization Problem for the Mechanical System

If matrix (3) is observed, it is possible to notice that this matrix is a block matrix and thus represents, as already observed, an internally decoupled system. Considering that, together with the preselecting matrix $\mathbf{T}_\theta$ calculated above, it is possible to consider a centralized model predictive controller as follows. Considering $k = nT_s$ and $k+1 = (n+1)T_s$, where T_s represents the sampling time and $n \in \mathbb{N}$, then the following expression can be obtained:

$$\begin{aligned} \mathbf{x}(k+1) &= \mathbf{A}_k\mathbf{x}(k) + \mathbf{B}_k u_{mpc}(k) \\ y_h(k) &= \mathbf{H}_k\mathbf{x}(k), \end{aligned} \tag{34}$$

where $\mathbf{A}_k$ represents the discretization of the block matrix of definition (4) of matrix $\mathbf{A}$, matrix $\mathbf{B}_k$ represents in the same way the discretization of each block matrix of definition (33) of matrix $\mathbf{B}$, $\mathbf{x}(k)$ is the corresponding discrete part of the state variable and matrix $\mathbf{H}_k$ is the output matrix which determines the positions (x, y and θ positions). The length of the prediction horizon which is considered equal to four samples and the following prediction expressions are obtained:

$$\hat{y}_h(k+1/k) = \mathbf{H}_k\mathbf{A}_k\mathbf{x}(k/k) + \mathbf{H}_k\mathbf{B}_k u_{mpc}(k) \tag{35}$$

$$\hat{y}_h(k+2/k) = \mathbf{H}_k\mathbf{A}_k^2\mathbf{x}(k/k) + \mathbf{H}_k\mathbf{A}_k\mathbf{B}_k u_{mpc}(k) + \mathbf{H}_k\mathbf{B}_k u_{mpc}(k+1), \tag{36}$$

and thus

$$\begin{aligned} \hat{y}_h(k+3/k) = {} & \mathbf{H}_k\mathbf{A}_k^3\mathbf{x}(k/k) + \mathbf{H}_k\mathbf{A}_k^2\mathbf{B}_k u_{mpc}(k) \\ & + \mathbf{H}_k\mathbf{A}_k\mathbf{B}_k u_{mpc}(k+1) + \mathbf{H}_k\mathbf{B}_k u_{mpc}(k+2), \end{aligned} \tag{37}$$

$$\begin{aligned} \hat{y}_h(k+4/k) = {} & \mathbf{H}_k\mathbf{A}_k^4\mathbf{x}(k/k) + \mathbf{H}_k\mathbf{A}_k^3\mathbf{B}_k u_{mpc}(k) + \\ & \mathbf{H}_k\mathbf{A}_k^2\mathbf{B}_k u_{mpc}(k+1) + \mathbf{H}_k\mathbf{A}_k\mathbf{B}_k u_{mpc}(k+2) + \\ & \mathbf{H}_k\mathbf{B}_k u_{mpc}(k+3), \end{aligned} \tag{38}$$

It is straightforward to show that the following vectorial expression holds:

$$\hat{\mathbf{Y}}_h(k) = \mathbf{G}_p\mathbf{x}(k) + \mathbf{F}_{1p}\mathbf{U}_{mpc}(k), \tag{39}$$

where

$$\hat{\mathbf{Y}}_h(k) = \begin{bmatrix} \hat{y}_h(k+1/k) \\ \hat{y}_h(k+2/k) \\ \hat{y}_h(k+3/k) \\ \hat{y}_h(k+4/k) \end{bmatrix}, \qquad \mathbf{U}_{mpc}(k) = \begin{bmatrix} u_{mpc}(k) \\ u_{mpc}(k+1) \\ u_{mpc}(k+2) \\ u_{mpc}(k+3) \end{bmatrix} \tag{40}$$

and matrices $\mathbf{G}_p$, $\mathbf{F}_{1p}$ are given here below:

$$\mathbf{G}_p = \begin{bmatrix} \mathbf{H}_k\mathbf{A}_k \\ \mathbf{H}_k\mathbf{A}_k^2 \\ \mathbf{H}_k\mathbf{A}_k^3 \\ \mathbf{H}_k\mathbf{A}_k^4 \end{bmatrix}, \quad \mathbf{F}_{1p} = \begin{bmatrix} \mathbf{H}_k\mathbf{B}_k & \mathbf{0} & \mathbf{0} & \mathbf{0} \\ \mathbf{H}_k\mathbf{A}_k\mathbf{B}_k & \mathbf{H}_k\mathbf{B}_k & \mathbf{0} & \mathbf{0} \\ \mathbf{H}_k\sum_{i=1}^{2}\mathbf{A}_k^i\mathbf{B}_k & \mathbf{H}_k\mathbf{A}_k\mathbf{B}_k & \mathbf{H}_k\mathbf{B}_k & \mathbf{0} \\ \mathbf{H}_k\sum_{i=1}^{3}\mathbf{A}_k^i\mathbf{B}_k & \mathbf{H}_k\sum_{i=1}^{2}\mathbf{A}_k^i\mathbf{B}_k & \mathbf{H}_k\mathbf{A}_k\mathbf{B}_k & \mathbf{H}_k\mathbf{B}_k \end{bmatrix}. \tag{41}$$

If the following performance criterion is assumed,

$$J = \frac{1}{2}\sum_{j=1}^{N}\Big(\Omega_d(k+j) - \hat{\Omega}(k+j)\Big)^{\mathrm{T}}\mathbf{Q}_p\Big(\Omega_d(k+j) - \hat{\Omega}(k+j)\Big) + \sum_{j=1}^{N-1}\Big(u_{mpc}(k+j)\Big)^{\mathrm{T}}\mathbf{R}_p u_{mpc}(k+j), \quad (42)$$

where $\Omega_d(k+j)$, $j = 1, 2, \ldots, N$ is the x, y position and θ reference trajectories and $\Omega(k+j)$ are the predicted ones. N is the prediction horizon, and $\mathbf{Q}_p$ and $\mathbf{R}_p$ are non-negative definite matrices. To be more precise:

$$\mathbf{H}_k = \begin{bmatrix} 1 & 0 & 0 & 0 & 0 & 0 \\ 0 & 0 & 1 & 0 & 0 & 0 \\ 0 & 0 & 0 & 0 & 1 & 0 \end{bmatrix} \quad (43)$$

is the matrix which selects the x, y and θ positions. Vector

$$\hat{\Omega}(k+j) = \hat{\mathbf{Y}}_h(k+j) = \begin{bmatrix} \hat{x}(k+j) \\ \hat{y}(k+j) \\ \hat{\theta}(k+j) \end{bmatrix} \quad (44)$$

represents the predicted three outputs structured in a column vector format. In a similar way vector

$$\Omega_d(k+j) = \mathbf{Y}_{h_d}(k+j) = \begin{bmatrix} \hat{x}_d(k+j) \\ \hat{y}_d(k+j) \\ \hat{\theta}_d(k+j) \end{bmatrix} \quad (45)$$

represents the desired future three outputs structured in a column vector format.

The solution minimizing performance index (42) may then be obtained by solving

$$\frac{\partial J}{\partial \mathbf{U}_{mpc}} = 0. \quad (46)$$

A direct computation of the vectorial optimal solution for the optimization problem defined above is as follows:

$$\tilde{\mathbf{U}}_{mpc} = (\mathbf{F}_{1p}^{\mathrm{T}}\mathbf{Q}_p\mathbf{F}_{1p} + \mathbf{R}_p)^{-1}\Big(\mathbf{F}_{1p}^{\mathrm{T}}\mathbf{Q}_p\big(\Omega_{d_p}(k) - \mathbf{G}_p x(k)\big)\Big), \quad (47)$$

where $\Omega_{d_p}(k)$ is the desired output column vector. The component at time "k" of the optimal vector of Eq. (47) is to be considered. Further details concerning this MPC can be found in [12].

5. Simulation Results

Figure 3 shows the general setup used for simulation and validation. Using a centralized MPC-controller with the geometric decoupling approach acting as a pre-control, different

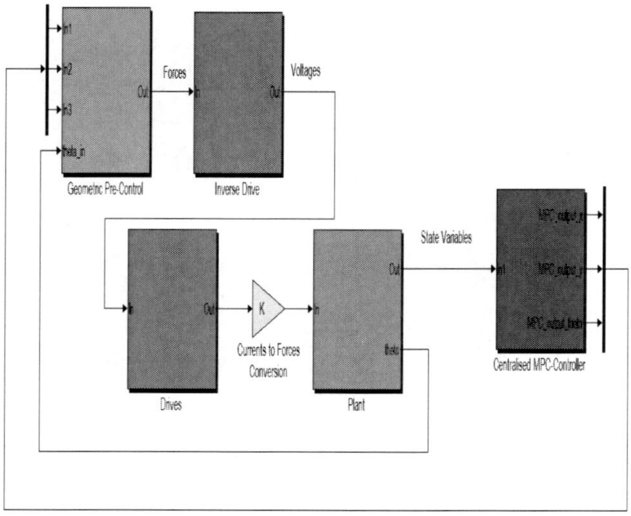

Figure 3. Block diagram of the control system structure.

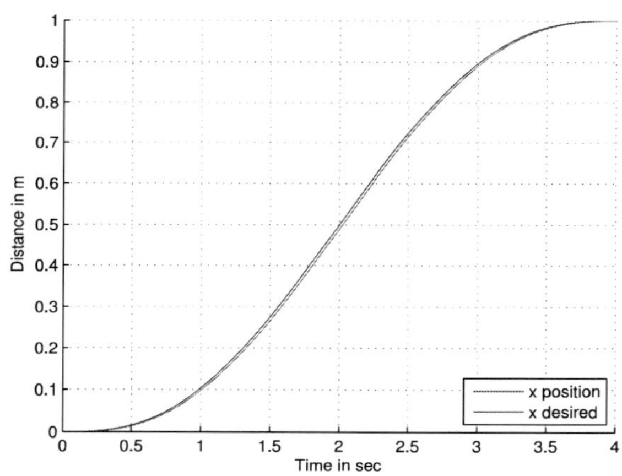

Figure 4. X-Trajectory of the Robotino.

trajectories can be tracked. MPC accepts trajectories in horizontal, vertical and angular directions and attempts to direct the Robotino only in the specified direction. Therefore, it is possible to move the Robotino along the x-axis without a rotation, for example. In the case presented here the robot was to move in the x-direction along a predetermined trajectory. To prevent any strains on the mechanical parts, as well as potential slipping, the trajectory has the form of a sixth-order polynomial. As seen in Fig. 4, the tracking works very well, while no movement in the y-direction or rotational movement occurs. Fig. 5 shows the outcome; movement here is very small and at this time contributed to numerical errors in the simulation.

Figure 6 shows the output of the MPC algorithm, the desired forces in x, y and angle direction. These results were obtained keeping the drive currents within the boundaries of

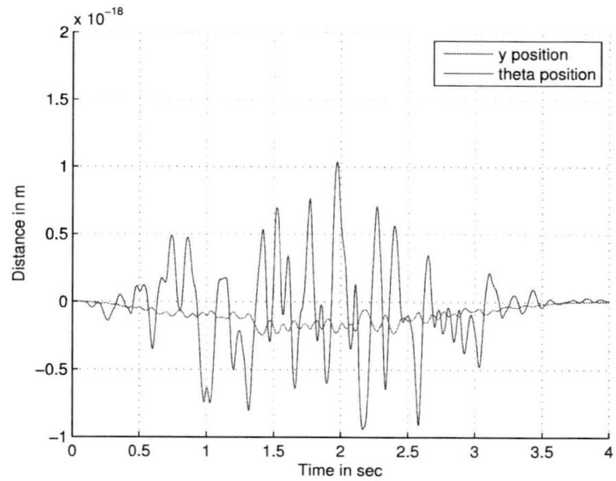

Figure 5. Y and Theta movement.

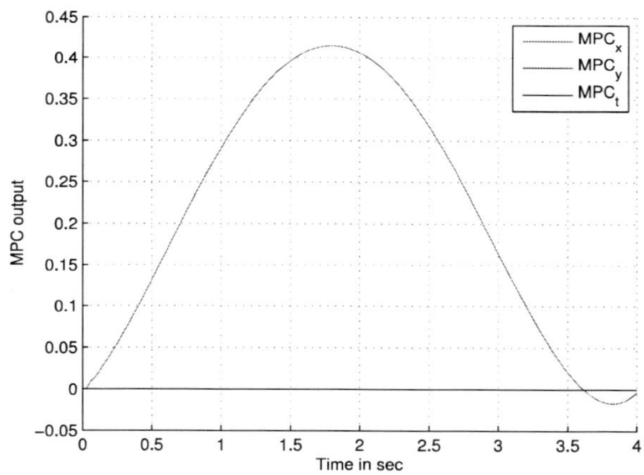

Figure 6. Output of MPC (MPC_y and MPC_t both equal zero).

the drives' nominal currents, $i(t) \in [-0.9, 0.9]$ Ampere, shown in Fig. 7.

6. Conclusion

This contribution presents a procedure in order to obtain the non-interacting functional controllability between horizontal, vertical and angular motions using a geometric approach. Here, the functional non-interaction problem is investigated. To achieve non-interaction a feedback control law is needed, together with a feed-forward regulator. The relevance of the motion functional controllability to the Robotino control is justified by the necessity of very fast and very precise loops of acceleration control counteracting errors in displacements caused by possible disturbance actions. Simulation results are shown to indicate the potentiality of this approach.

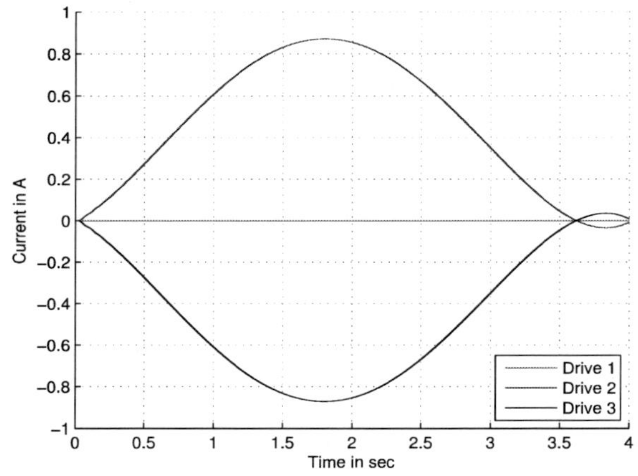

Figure 7. Currents of the three drives acting the three wheels.

References

[1] G. Basile and G. Marro. *Controlled and conditioned invariants in linear system theory*. Prentice Hall, New Jersey, 1992.

[2] A. Isidori. *Nonlinear control Systems: an introduction*. Springer-Verlag, Berlin, 1989.

[3] W.M. Wonham. *Linear multivariable control: a geometric approach*. Springer-Verlag, New York, 1979.

[4] P. Mercorelli and D. Prattichizzo. A geometric procedure for robust decoupling control of contact forces in robotic manipulation. *Kybernetika*, 39(4):433–455, 2003.

[5] D. Prattichizzo and P. Mercorelli. On some geometric control properties of active suspension systems. *Kybernetika*, 36(5):549–570, 2000.

[6] P. Mercorelli. Invariant subspace for grasping internal forces and non-interacting force-motion control in robotic manipulation. *Kybernetika*, 48(6):1229–1249, 2012.

[7] P. Mercorelli. Geometric structures for the parameterization of non-interacting dynamics for multi-body mechanisms. *International Journal of Pure and Applied Mathematics-IJPAM*, 59(3):257–273, 2010.

[8] Y. Yamamoto and X. Yun. Effect of the dynamic interaction on coordinated control of mobile manipulators. *IEEE Transactions on Robotics and Automation*, 12(5):816–824, 1996.

[9] G. Basile and G. Marro. A state space approach to non-interacting controls. *Ricerche di Automatica*, 1(1):68–77, 1970.

[10] W.M. Wonham and A.S. Morse. Decoupling and pole assignment in linear multivariable systems: a geometric approach. *SIAM J. Control*, 8(1):1–18, 1970.

[11] A.S. Morse and W.M. Wonham. Decoupling and pole assignment by dynamic compensation. *SIAM J. Control*, 8(1):317–337, 1970.

[12] S. Huang, K.K. Tan, and T.H. Lee. *Applied Predictive Control.* Springer-Verlag London, Printed in Great Britain, 2002.

In: Model Predictive Control
Editor: Corrine Wade, pp. 77-98
ISBN: 978-1-63463-859-3
© 2015 Nova Science Publishers, Inc.

Chapter 4

ROBUST MODEL PREDICTIVE CONTROL OF HAMMERSTEIN SYSTEMS

***Silvina Inés Biagiola*[*] *and José Luis Figueroa*[†]**
Departamento de Ingeniería Eléctrica y de Computadoras,
IIIE (UNS-CONICET), Bahía Blanca, Argentina

Abstract

There are very few controller design techniques that can be proven to stabilize processes in the presence of nonlinearities and constraints. Model predictive control (MPC) is one of these techniques. For this reason, there has been much interest in nonlinear model-based control within the process engineering community. A critical step in the application of these methods is the development of a suitable model for the process dynamics. In this sense, block-oriented models have proved to be useful as simple nonlinear models for a vast number of applications. They are described as a cascade of linear dynamic and nonlinear static blocks. They have emerged as an appealing proposal due to their simplicity and the property of being valid over a larger operating region than a linear time invariant (LTI) model. A typical block-oriented model found in the literature is the Hammerstein model. In this structure a nonlinear memoryless block is followed by a linear dynamics. A broad type of dynamic processes can be described by such representations consisting of these two simple elements usually referred to as subsystems. This chapter deals with robust control for uncertain Hammerstein models. The starting point for the controller design is a Hammerstein model which describes the systems dynamics in the presence of uncertainty. This model is employed to design a model based predictive controller. The mathematical problem involved in the development of the algorithm is stated in the context of Linear Matrix Inequalities (LMI) theory. The straightforward use of Hammerstein models for designing the Model Predictive controller would lead to a nonlinear optimization problem due to the static nonlinearity. From the point of view of the implementation, this could result in high computational complexity and be a very time-consuming process. This can be avoided by exploiting the structure of the Hammerstein model, which is a novel approach. This strategy developed in this chapter takes advantage of

[*]E-mail address: biagiola@uns.edu.ar
[†]E-mail address: figueroa@uns.edu.ar

the static nature of the nonlinearity which allows being transformed into polytopic representation and, therefore, to solve the control problem by focusing only in the linear dynamics. This formulation results in a simplified design procedure, because the original nonlinear Model Predictive Control problem turns into a linear one. At the end of the chapter, different simulation examples are presented to illustrate the controller design procedure.

Keywords: uncertain model, model predictive control, linear matrix inequalities, robust control, optimization

1. Introduction

Although industrial processes are inherently nonlinear, many contributions for controller design for those plants are based on the assumption of a linear model of the system. However, in some cases it is difficult to represent a given process using a linear model. This is the case, for example, when the system is highly nonlinear and the operating point changes along a wide region.

There are very few controller design techniques that can be proven to stabilize processes in the presence of nonlinearities and constraints. Model predictive control (MPC) is one of these techniques. For this reason, there has been much interest in nonlinear model-based control within the process engineering community. A critical step in the application of these methods is the development of a suitable model of the process dynamics, Sjöberg et al. {1995} described several approaches for model development. A special class of non-linear models are the so-called block oriented nonlinear models in which a linear time invariant (LTI) dynamic block is preceded and/or followed by a static non-linearity (Biagiola and Figueroa {2009}, and references therein). When the linear dynamic block is preceded by a static input nonlinearity (see Figure 1) the model is referred to as a Hammerstein model (HM), which is one of the simplest and most popular members of the general family of block-oriented nonlinear models.

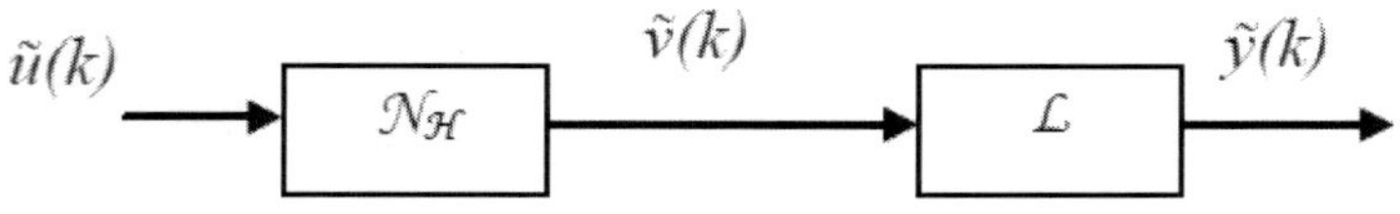

Figure 1. Schematic representation of a Hammerstein model. The input to the plant is denoted by $\tilde{u}$, which is mapped to $\tilde{v}$ by the static nonlinearity $\mathcal{N}_{\mathcal{H}}$. Signal $\tilde{v}$ is the input to the linear dynamic block $\mathcal{L}$, and it is mapped to the plant output $\tilde{y}$.

Although HMs only represent a small subclass of the totality of nonlinear models, they have emerged as an appealing proposal due to their simplicity and the property of being valid over a larger operating region than a LTI model. Hammerstein models have been useful for modeling many nonlinear processes, for example distillation columns (Eskinat et al. {1991}; Sentoni et al. {1996}; Pearson and Pottmann {2000}; Gómez and Baeyens {2004};

Biagiola and Figueroa {2011a}), pH neutralization processes (Fruzzetti et al. {1997}; Patwardhan et al. {2003}), heat exchanger behavior (Eskinat et al. {1991}), a pellet cooling process (Pomerleau et al. {2003}) and biological systems (Hunter and Korenberg {1986}; Durfee and Maclean {1989}; Westwick and Kearney {2001}; Dempsey and Westwick {2004}; Le et al. {2010}). One typical process that favors the selection of a Hammerstein model is the presence of an actuator nonlinearity like a smooth saturation (Bloemen et al. {2001}). HMs models have applications in many engineering problems and, therefore, identification of these mathematical representations has been an active research area for many years (Janczak {2005} and references therein, Bai and Li {2010}, Chen et al. {2012}; Mattsson and Wigren {2014}).

On the other hand, control of Hammerstein systems is a relatively more recent problem, that started to be dealt with later in the literature. Moreover, most of the algorithms developed for HMs control, are oriented to "nominal" models. Dolanc and Strmčnik {2008} proposed a nonlinear controller for a piecewise-linear Hammerstein model, and the design was based on the principles of a classic linear pole placement controller. Patwardhan et al. {2003} proposed a constrained nonlinear model-predictive control strategy. Their approach is based on the partial least square (PLS) framework. They concluded that PLS-based HMs for control involves solution of a nonlinear program as a result of the constraint mapping. One of the most referenced works regarding HMs control is the one by Fruzzetti et al. {1997}. In this article, the closed-loop system representation is transformed into one in which the linear MPC controller is retained, and the inverse of the static nonlinearity is applied to this scheme.

A common factor among the above control proposals is that they make use of the advantageous (Hammerstein) model structure in order to design a controller algorithm which relaxes the computational effort.

Bloemen et al. {2001} introduced an improvement to the previous approaches, in the sense that their proposal includes the static nonlinearity model in the controller algorithm design. Their method transforms the nonlinearities into polytopic descriptions. In such a way, the procedure enables the use of robust linear MPC techniques for controlling Hammerstein systems, while convex optimization problem is retained. This problem formulation based on polytopic nonlinear description, becomes appropriate to design a controller for HMs with uncertainty in the static nonlinearity.

The presence of uncertainty is not a minor problem, and it deserves attention at the time of controller design. Nominal model identification, usually relies on some prior assumptions such as model structure and order, which are often unavailable or subject to uncertainties. Hence, the complexity and errors associated with such nominal models increase the difficulty of the control design task, and may lead to degradation of control performance. Therefore, robust identification of uncertain Hammerstein models appears as an appealing proposal, and some contributions on this subject can be found in Bloemen et al. {2001}, Biagiola and Figueroa {2009}, Biagiola and Figueroa {2011a}.

However, to the best of the authors knowledge, a robust approach to design controllers for this HMs under uncertainty in both the linear and nonlinear blocks has not been developed in the literature.

In this work a controller synthesis procedure is developed for uncertain HMs. A model predictive-based strategy is used for the controller design, recalling the methodology presented in Kothare et al. {1996} and later extended in Araújo and Oliveira {2009} to linear models represented by orthonormal basis functions (Biagiola and Figueroa {2011b}). The uncertainty in the whole nonlinear model is explicitly considered in the controller synthesis. This is achieved by the inclusion of uncertainty model in the algorithm design. It is assumed that the uncertainty in the linear block can be described following the "multi-model paradigm" introduced by Kothare et al. {1996}. Additionally, the nonlinear block uncertainty is analyzed as a conic-sector (Biagiola et al. {2004}) following a polytopic description (Bloemen et al. {2001}).

The goal is to design, at each time step, a state-feedback control law that minimizes a "worst-case" infinite horizon objective function, subject to possible constraints on the control input and/or plant output. Using linear matrix inequalities (LMIs), the problem of minimizing an upper bound on the "worst-case" objective function, subject to existing constraints, is reduced to a convex optimization one. Robustness properties of the proposed algorithm result from the Lyapunov-based controller design.

This chapter is organized in the following way. In Section 2, the mathematical description assumed for the uncertain Hammerstein model is introduced. This is the starting point for the control problem treated in Section 3, in which robust predictive controller synthesis for uncertain Hammerstein models is addressed. Two different simulation examples are presented in Section 4 to illustrate the implementation of the control algorithm. In Section 5 some comments about the proposed algorithm and the simulation results are presented. This contribution concludes with some final remarks in Section 6

2. Problem Formulation

2.1. Nominal Model Description

Hammerstein Models (HMs) consist of a process with nonlinear gain followed by a linear dynamics (Figure 1), and they can represent many of the nonlinearities commonly encountered in industrial processes.

The problem of nominal MPC for Hammerstein-Wiener models was studied by Bloemen et al. {2001}. The mathematical expression used in that paper for the case of pure Hammerstein systems is

$$\tilde{v}(k) = \mathcal{N}_{\mathcal{H}}(\tilde{u}(k)) \tag{1}$$

$$\tilde{x}(k+1) = \tilde{A}\tilde{x}(k) + \tilde{B}\tilde{v}(k), \tag{2}$$

$$\tilde{y}(k) = \tilde{C}\tilde{x}(k), \tag{3}$$

where $\tilde{u}(k) \in \Re^{n_u}$ is the control input, which is passed through the nonlinear mapping $\mathcal{N}_{\mathcal{H}}$ to give the input $\tilde{v}(k) \in \Re^{n_u}$. It should be remarked that the robust control approach proposed in this work does not impose to $\mathcal{N}_{\mathcal{H}}$ the constraint of being invertible. In this way, a wider set of Hammerstein models can be considered for the controller design procedure, in contrast with previous control algorithms (Fruzzetti et al. {1997}; Norquay et al. {1998}; Patwardhan et al. {2003}; Bloemen et al. {2001}).

The signal $\tilde{v}(k)$ is filtered by the linear dynamic block $(\tilde{A}, \tilde{B}, \tilde{C})$, where $\tilde{x}(k) \in \Re^{n_x}$ is the state of the plant, and $\tilde{y}(k) \in \Re^{n_y}$ is the plant output.

Rewriting the linear model in deviation variables, and taking into account a constant setpoint (y_{sp}), the following model is obtained (Araújo and Oliveira {2009}):

$$\begin{aligned} \begin{bmatrix} \Delta\tilde{x}(k+1) \\ \tilde{y}(k+1) - y_{sp} \end{bmatrix} &= \begin{bmatrix} \tilde{A} & 0 \\ \tilde{C}\tilde{A} & I \end{bmatrix} \begin{bmatrix} \Delta\tilde{x}(k) \\ \tilde{y}(k) - y_{sp} \end{bmatrix} + \begin{bmatrix} \tilde{B} \\ \tilde{C}\tilde{B} \end{bmatrix} \Delta\tilde{v}(k) \\ \begin{bmatrix} \tilde{y}(k) - y_{sp} \end{bmatrix} &= \begin{bmatrix} 0 & I \end{bmatrix} \begin{bmatrix} \Delta\tilde{x}(k) \\ \tilde{y}(k) - y_{sp} \end{bmatrix} \end{aligned} \quad (4)$$

where $\Delta = 1 - q^{-1}$ and q^{-1} is the delay operator. Further details of mathematical derivations of Eq. (4) are given in the Appendix.

Now, if the following definitions are made:

$$A = \begin{bmatrix} \tilde{A} & 0 \\ \tilde{C}\tilde{A} & I \end{bmatrix} \quad (5)$$

$$B = \begin{bmatrix} \tilde{B} \\ \tilde{C}\tilde{B} \end{bmatrix} \quad (6)$$

$$C = \begin{bmatrix} 0 & I \end{bmatrix} \quad (7)$$

$$x = \begin{bmatrix} \Delta\tilde{x}(k) \\ \tilde{y}(k) - y_{sp} \end{bmatrix} \quad (8)$$

$$v(k) = \Delta\tilde{v}(k) \quad (9)$$

Then, the complete linear model is described as

$$x(k+1) = Ax(k) + Bv(k), \quad (10)$$

$$y(k) = Cx(k). \quad (11)$$

The Hammerstein model is completed with the nonlinear description of Eqs. (1-3) where

$$\tilde{v}(k) = \tilde{v}(k-1) + v(k), \quad (12)$$

$$\tilde{u}(k) = \tilde{u}(k-1) + u(k). \quad (13)$$

Note that $u(k) = \Delta\tilde{u}(k)$, the incremental input to the Hammerstein structure, is the manipulated variable to be calculated by the MPC algorithm.

2.2. Uncertain Model Characterization

In the previous subsection, the nominal model was introduced. However, the goal is to develop an effective control strategy for Hammerstein models with uncertainty.

An exact plant model with same initial conditions and inputs can identically reproduce the outputs of the plant. However, the physics of the systems may not be completely understood in many applications. Therefore, in those cases an exact mathematical model of the process could be not only costly but impossible to obtain as well. Moreover, there are several systems that are highly nonlinear, time-varying, and infinite dimensional (e.g., many biological processes). Even if the exact plant model is available, this plant model cannot be used directly for the purposes of simulation or controller design because the model complexity usually turns the control problem intractable. The practical infeasibility to obtain reliable models motivates the need to consider the case where the nonlinear systems are uncertain, and to develop ad hoc models for both simulation and control. Therefore, derivation of a mathematical representation able to model uncertain MIMO Hammerstein systems is one of the main objectives. For this purpose, the nominal model developed in the previous section will be required.

Therefore, the first step is to provide a characterization for the uncertainty present in the model. This is a relevant definition because the robust controller design approach must be suitable to deal with the assumed uncertainty model.

In what follows it will be assumed that the uncertainty in the linear dynamics can be represented by a polytopic model or "multi-model" paradigm described in the seminal work by Kothare et al. {1996}.

The linear dynamics in Eq. (2-3) is such that $[\tilde{A} \quad \tilde{B}] \in \tilde{\Omega}$

$$\tilde{\Omega} = Co\{[\tilde{A}_1 \quad \tilde{B}_1], [\tilde{A}_2 \quad \tilde{B}_2], \ldots, [\tilde{A}_L \quad \tilde{B}_L]\} \tag{14}$$

where Co means convex hull. In other words, if $[\tilde{A} \quad \tilde{B}] \in \tilde{\Omega}$ then, for some $\tilde{\lambda}_i \geq 0$; $i = 1, \ldots, L$ with $\sum \tilde{\lambda}_i = 1$ we have

$$[\tilde{A} \quad \tilde{B}] = \sum_{i=1}^{L} \tilde{\lambda}_i [\tilde{A}_i \quad \tilde{B}_i]. \tag{15}$$

Then, the available knowledge for the controller design about the uncertainty in the linear block are the values of the L sets $[\tilde{A}_i \quad \tilde{B}_i]$. Note that this information can be straightforwardly transferred to the model description in Eqs. (10-11) and results in the time-variant sets $[A \quad B] \in \Omega$ where the redefined convex hull is

$$\Omega = Co\{[A_1 \quad B_1], [A_2 \quad B_2], \ldots, [A_L \quad B_L]\} \tag{16}$$

Now, it is necessary to define an uncertainty model for the static nonlinearity. For this purpose, the approach followed by Bloemen et al. {2001} is here adopted. That means the uncertain behavior of the nonlinearity $\mathcal{N}_{\mathcal{H}}$ is assumed to be represented by a polytopic description which is linear in u. Basically, sector bounds (Biagiola et al. {2004}) of the nonlinear function are determined (which are valid in the operating region) and the following inequalities arise

$$G_{min} \cdot u(k) \leq v(k) \leq G_{max} \cdot u(k) \tag{17}$$

which involves the uncertain nonlinear functions in $\mathcal{N}_{\mathcal{H}}$ pass the origin (Bloemen et al. {2001}). In other words,

$$v(k) = G(u(k)).u(k) \tag{18}$$

with

$$G(u(k)) \in \Omega_G = Co\{G_{min}, G_{max}\} \tag{19}$$

Note that the equations herein derived to cope with uncertainty together with the model structure developed in subsection 2.1 provide an uncertain Hammerstein model description. This can be interpreted as an appealing result, which is specially oriented for subsequent robust control design. The developed model allows dealing with uncertainty both in the linear and the nonlinear blocks. The modeling approach makes use of polytopic descriptions for both convex hulls, Ω and Ω_G.

3. Robust MPC Algorithm Design for Hammerstein Systems

Based on the uncertain model described in the previous section, an algorithm must be synthesized for controlling this kind of uncertain Hammerstein models. The approach herein followed deals with a robust model predictive control strategy.

According to this problem formulation, the unconstrained RMPC performance objective can be possed as

$$\min_{u(k+i|k),i=0,1,\ldots,m} \quad \max_{\substack{[A \;\; B] \;\in \Omega \\ G \;\in \Omega_G}} \quad J_\infty(k) \tag{20}$$

and the control goal is to minimize the quadratic performance index

$$J_\infty(k) = \sum_{i=0}^{\infty} \{x(k+i|k)^T \, Q_x \, x(k+i|k) + u(k+i|k)^T \, R_u \, u(k+i|k)\} \tag{21}$$

where the weights Q_x and R_u in (21) are diagonal matrices in $\Re^{n_x \times n_x}$ and $\Re^{n_u \times n_u}$, respectively; and $x(k+i|k)$ and $u(k+i|k)$ are the state and control move respectively, at time $k+i$, computed based on measurements at time k[1]. In particular, $x(k|k)$ is the measured state at time k and $u(k|k)$ is the control move to be implemented at time k. It is assumed that there is no control action after time $k+m-1$ (i.e. $u(k+i|k) = 0$ for $i \geq m$) where m is the control horizon. Due to the presence of $\mathcal{N}_{\mathcal{H}}$, this problem results in a nonlinear one.

The goal is to design, at each time step, a state-feedback control law that minimizes a worst-case infinite horizon objective function.

Note that Eq. (20) stands for a "min-max" problem. The maximization is over the set of possible plants, and corresponds to choosing a plant which, if used as model for predictions, would lead to the worst case value of $J_\infty(k)$ among all plants in $\Omega \bigcup \Omega_G$. This worst case value is then minimized over the present and future control moves $u(k+i|k)$, $i = 0, 1, \cdots, m$.

In this RMPC design, an infinite prediction horizon is considered. The reason for this is the infinite horizon enables to guarantee nominal closed loop stability (Mayne et al. {2000}; Camacho and Alba {2004}).

To solve this problem, an upper bound on $J_\infty(k)$ can be developed in the form of

$$J_\infty(k) \leq V(x(k|k)) \tag{22}$$

[1]It should be remarked that $x(k+i|k)$ (see Eq. (8)) and $u(k+i|k)$ (see Eq. (13)) stand for the incremental values in the state and manipulated variables, respectively, and not their absolute values.

where $V(x) = x^T P x$ with $P > 0$. Thus

$$\max_{\substack{[A\ \ B]\ \in \Omega \\ G\ \in \Omega_G}} J_\infty(k) \leq V(x(k|k)). \tag{23}$$

This expression gives an upper bound on the robust performance objective. Thus, following Kothare et al. {1996}, the objective of the MPC can be redefined to compute, at each sample time, a constant control law $u(k+i|k) = Fx(k+i|k)$ to minimize this upper bound $V(x(k|k))$.

As is usual in MPC, the first input $u(k|k)$ is implemented. At the next time, the variable $x(k+1)$ is adapted[2], and the optimization is repeated to compute a new F (which can be interpreted as an adaptive gain). The solution of the optimization problem is obtained on the basis of the following theorem (Kothare et al. {1996}).

Theorem 1. *Let $x(k) = x(k|k)$ be the state of the uncertain system (10-11). Then the state feedback matrix F in the control law $u(k+i|k) = Fx(k+i|k)$, $i \geq 0$ that minimizes the upper bound $V(x(k|k))$ on the robust performance objective function at sampling time k is given by*

$$F = YQ^{-1}, \tag{24}$$

where $Q > 0$ and Y are obtained from the solution (if it exists) of the following linear objective minimization problem with LMI constraints,

$$\min_{\gamma,Q,Y} \gamma \tag{25}$$

subject to

$$\begin{bmatrix} 1 & x(k|k)^T \\ x(k|k) & Q \end{bmatrix} \geq 0 \tag{26}$$

and

$$\begin{bmatrix} Q & QA_j^T + Y^T(B_jG_l)^T & QQ_x^{1/2} & Y^T R_u^{1/2} \\ A_jQ + (B_jG_l)Y & Q & 0 & 0 \\ QQ_x^{1/2} & 0 & \gamma I & 0 \\ R_u^{1/2} Y & 0 & 0 & \gamma I \end{bmatrix} \geq 0, \tag{27}$$

$$j = 1, 2, \ldots, L; l = min, max.$$

Proof. Taking into account Eq. (18), the model presented in Eqs. (10-11) can be rewritten as follows

$$x(k+1) = Ax(k) + BGu(k), \tag{28}$$

$$y(k) = Cx(k). \tag{29}$$

[2]Note that in this formulation, the only information required from the process is the controlled output $y(k)$ and not the whole state vector. This can be checked by inspection of the definition of x in Eq. (8), since it is formed by the increment of the state, the set point and the measured output. Therefore, an output feedback approach is herein proposed. In this way, the process states values are not necessary, which contrasts with the commonly used state space approaches for MPC.

If the bounds on G assumed in Eq. (17) are considered, then the polytope Ω described in Eq. (16) for the (v, y) input-output linear model can be redefined for the (u, y) input-output Hammerstein model. Therefore, the resultant polytope $\Omega_{uy} = \Omega \bigcup \Omega_G$ is

$$\begin{aligned} \Omega_{uy} &= \Omega \bigcup \Omega_G \\ &= Co\{[A_1 \;\; B_1 G_{min}], [A_1 \;\; B_1 G_{max}], \ldots, [A_L \;\; B_L G_{min}], [A_L \;\; B_L G_{max}]\} \end{aligned} \tag{30}$$

Note that the number of the vertices in this polytope duplicates the number of vertices of Ω. The LMIs formulated in Eq. (27) arise from this polytope. From now on, the proof continues as in Kothare et al. {1996}.

Note that the robust predictive control algorithm herein proposed for Hammerstein systems is inherently associated to a min-max optimization problem. The general formulation is given by Eq. 20, which means that the "minimization" is associated to the attempt to achieve minimum control cost, and the "maximization" of the functional (J) involves the attempt to design a controller being able to treat even with the worst-case scenario. This disadvantageous situation, denoted as worst-case, is due to the uncertainty present in the system. Although it could hardly take place, the control action is calculated in such a way that the controller is "prepared" for this improbable situation. The statement in Eq. 20, which is not computationally tractable in such a form (Kothare et al. {1996}) can be reformulated as shown in Eq. 23, which involves solving the same optimization problem but in a more convenient form. In other words, the control purpose stated in Eq. 20 is retained in Eq. 23 but in a more suitable mathematical approach which allows solving the problem through conventional LMI methodology. In Eq. 23 the "minimization" task is not as obvious as in Eq. 20, and it consists in finding the input sequence $u(k)$ which minimizes the upper bound $V(x(k|k))$.

The formulation in theorem 1. allows solving the optimization problem in Eq. 23 for a constant control law $u(k+i|k) = Fx(k+i|k)$. The control movements are calculated to minimize the upper bound of $V(x(k|k))$. As mentioned in Kothare et al. {1996} and references therein, the formulation in Eqs. (24-27) can be derived from the results for quadratic stabilization of uncertain polytopic continuous-time systems and their extension to the discrete-time case. It is clear that the problem in Eq. 25 consists in a linear objective minimization problem, which written in this form belongs to the set of the so-called eigenvalue problems (EVP) (Boyd et al. {1994}). The reformulation of the optimization problem is possible due to the use of LMI theory which generates a linear objective minimization problem with three optimization variables: γ, Q and Y. Further details on these variables generation can be read in the Appendix section of Kothare et al. {1996}.

The previous unconstrained optimization problem could be extended to consider constraints on the manipulated variables and/or on the process outputs. If constraints on the control variables are set by $\|u(k+i|k)\|_2 \le u_{max}$, for $i \ge 0$, the following LMI should be included as a constraint in problem (25-27)

$$\begin{bmatrix} u_{max}^2 I & Y \\ Y^T & Q \end{bmatrix} \ge 0. \tag{31}$$

In the presence of output constraints of the form

$$\max_{\substack{[A \ B] \in \Omega \\ G \in \Omega_G}} \|y(k+i|k)\|_2 \leq y_{max} \tag{32}$$

the LMI constraint to be included in (25-27) is

$$\begin{bmatrix} Q & (A_jQ + (B_jG_l)Y)^T C^T \\ C(A_jQ + (B_jG_l)Y) & y_{max}^2 I \end{bmatrix} \geq 0, j = 1, 2, \ldots, L; l = 1, 2. \tag{33}$$

It should be noted that thought this work makes use of the tool given in Kothare et al. {1996} for the statement of the optimization problem, the proposal herein presented is by no means restricted to that approach. For example, other formulations such the one by Veselý et al. {2010}, could be used to accomplish the optimization. That paper, which can be included in the group of works dealing with explicit MPC approaches (Alessio and Bemporad {2009}) also focuses on finding an LMI feasible solution.

4. Example

The goal of this section is to illustrate the implementation of the robust model-predictive controller previously described. The algorithm performance is shown by means of two simulation examples: a SISO Hammerstein model is first considered and, secondly, a model predictive controller is developed for a MIMO Hammerstein model.

4.1. SISO Hammerstein Model

The aim of the first example is to show the controller design proposal for a SISO system. In this way, and to avoid further complexity, a simple case is developed to exemplify the methodology and implementation details.

The model herein considered is adapted from Example 2 presented in Kothare et al. {1996}. A smooth saturation for the manipulated variables is applied to the discrete-time state-space equations considered for the linear dynamics presented in Kothare et al. {1996}. In this way, the static nonlinearity is of a type often encountered in practice Bloemen et al. {2001}. Therefore, the mathematical expression for this Hammerstein model is (see Fig. 1):

$$v_{nom}(k) = (e^u - 1)/(e^u + 1) \tag{34}$$

$$x(k+1) = \begin{bmatrix} 1 & 0 & 0.1 & 0 \\ 0 & 0 & 0 & 0.1 \\ -K & K & 1 & 1 \\ K & -K & 0 & 1 \end{bmatrix} x(k) + \begin{bmatrix} 0 \\ 0 \\ 0.1 \\ 0 \end{bmatrix} v(k), \tag{35}$$

$$y(k) = \begin{bmatrix} 0 & 1 & 0 & 0 \end{bmatrix} x(k),$$

where the parameter K is such that $K \in \{0.5, 2\}$. The polytopic representation (conic sectors) for v is shown in Fig. 2. For this example, the conic sectors are defined by $G_{min} = 0.241$ and $G_{max} = 0.500$. The weighting matrices in the performance index (Eq. 21) are: $Q_x = I$ and $R_u = 0.05$.

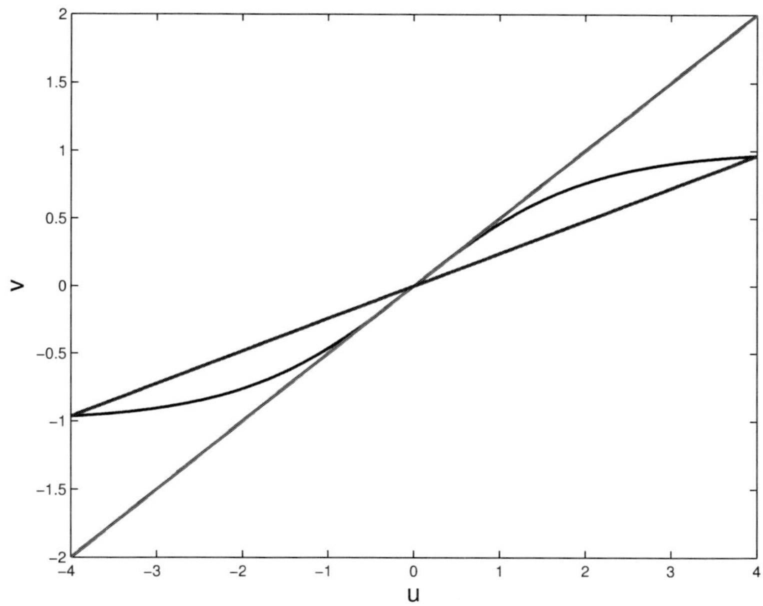

Figure 2. Polytopic representation for the input nonlinearity. Solid line: nonlinearity (Eq.34). Blue and red lines: sector bounds of the nonlinearity (in the operating region).

To evaluate the robust model predictive controller, three different situations were considered. Figure 3 shows the performance achieved with the proposed robust controller for three different plants: $K = 0.5, 1.25$ and 2. Note that in order to represent the nominal model, the parameter $K = 1.25$ was replaced in Eq. 35. As regards the formulation of the LMI problem, the dimensions of the optimization variables are $Q \in \Re^{5\times 5}$, $Y \in \Re^{5\times 1}$ and $\gamma \in \Re$.

The matrices and vectors involved in the LMI problem are:

$$A_j = \begin{bmatrix} 1 & 0 & 0.1 & 0 & 0 \\ 0 & 1 & 0 & 0.1 & 0 \\ -K_j & K_j & 1 & 0 & 0 \\ K_j & -K_j & 0 & 1 & 0 \\ 0 & 1 & 0 & 0 & 1 \end{bmatrix}, j = 1, 2; \tag{36}$$

with $K_1 = 0.5$ and $K_2 = 2$ (i.e., the extreme values of K) and

$$B_l = \begin{bmatrix} 0 \\ 0 \\ 0.1G_l \\ 0 \\ 0 \end{bmatrix}; C = \begin{bmatrix} 0 & 0 & 0 & 0 & 0 \end{bmatrix}; D = 0, \tag{37}$$

with $l = min, max$. In this way, the inequality given in Eq. 26 involves a 6×6 matrix. Such inequality is the responsible of adapting the state vector (x) every sampling time. This

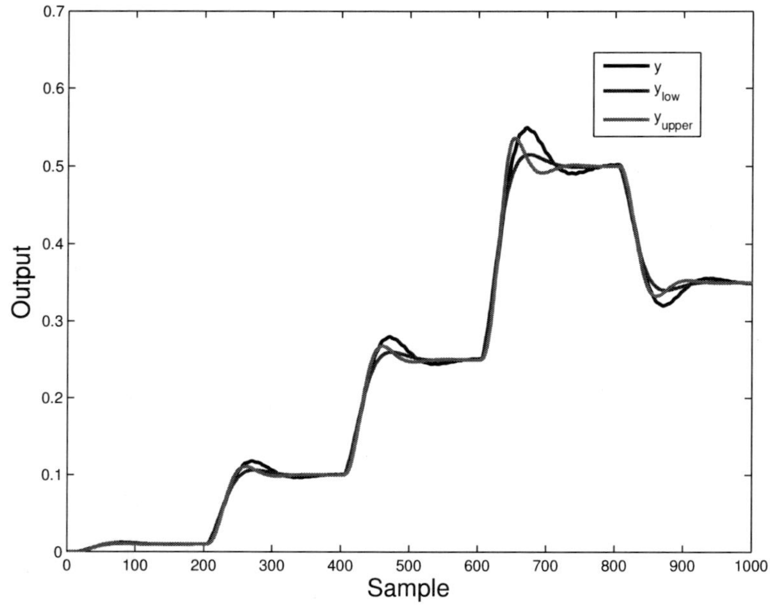

Figure 3. Control performance for the uncertain Hammerstein SISO model. Three different possible plants were considered (and controlled with the robust MPC algorithm): those in which the uncertain model parameter is $K = 1.25$ (black line), $K = 0.5$ (blue line) and $K = 2$ (red line).

vector is formed by the four model states plus the difference between the measured output and the setpoint (Eq. 8).

The remaining matrix inequalities are defined a priori based on $A_j(j = 1, 2)$ and $B_l(l = min, max)$. Therefore, a total amount of four sets of matrices are obtained.

The line plotted in black stands for the performance achieved when the "real" system is assumed to be such that the variable parameter is $K = 1.25$. The output of this "real" model is fed to the robust MPC algorithm, which calculates the control movements depicted in Fig. 4. Additional simulations were accomplished to evaluate the control performance. For this purpose, extreme values were given to K (i.e. 0.5 and 2). In these cases, the controlled outputs plotted in blue and red (see Fig. 3) were, respectively, obtained.

The MPC algorithm is also compared with the one introduced by Bloemen et al. {2001}. To accomplish this simulation the plant with $K = 1.25$ was considered. Both outputs are shown in Fig. 5. Both situations related to the limit values for K (i.e. $K = 0.5$ and $K = 2$) were also taken into consideration. However, the algorithm in Bloemen et al. {2001} led to an unstable closed-loop behavior and, therefore, it happens to be less robust than the algorithm herein introduced.

4.2. MIMO Hammerstein Model

To complete the illustration of the proposed method, a more complex case study is performed, with basis in the two input-two output model dealt with in Chan and Bao {2007}.

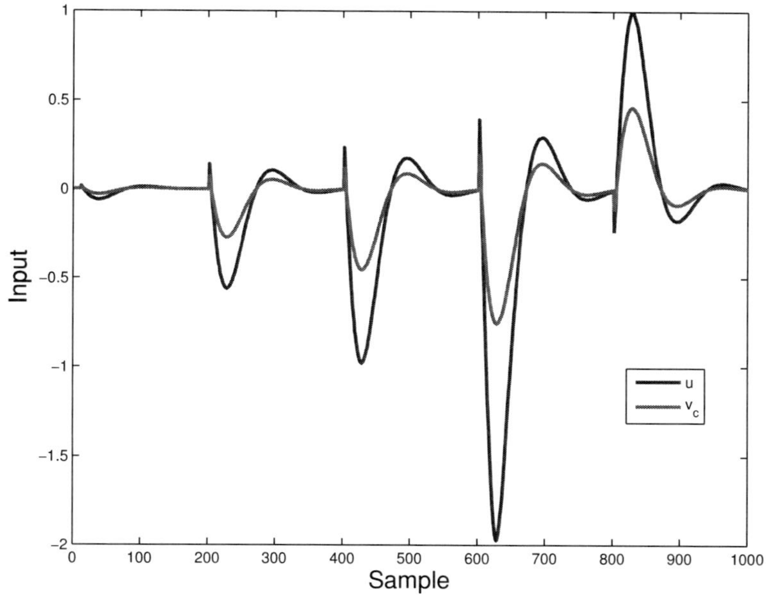

Figure 4. Control movements for the uncertain SISO model: robust controller output (blue) and calculated intermediate signal v_c (green).

The nominal Hammerstein system in the form of Eqs. (1-3) is as follows

$$\tilde{v}_1(\tilde{u}) = \tilde{u}_1 + 0.2\tilde{u}_1\tilde{u}_2 - 0.1\tilde{u}_1\tilde{u}_2^2 + \tilde{u}_2 \tag{38}$$

$$\tilde{v}_2(\tilde{u}) = -\tilde{u}_1 - 0.2\tilde{u}_1\tilde{u}_2 + 0.5\tilde{u}_1^2\tilde{u}_2 + 2\tilde{u}_2 \tag{39}$$

with

$$\tilde{v} = \begin{bmatrix} \tilde{v}_1 & \tilde{v}_2 \end{bmatrix}^T, \tag{40}$$

and

$$A = \begin{bmatrix} 0.3780 & 0.1331 & -0.1651 & -0.4715 \\ 0.2182 & 0.2275 & 0.5139 & -0.1475 \\ 0.4626 & -0.0954 & -0.0255 & 0.2939 \\ -0.0753 & -0.5538 & 0.1346 & -0.1260 \end{bmatrix}, \tag{41}$$

$$B = \begin{bmatrix} -10.6197 & 9.9973 \\ 11.5842 & -10.4204 \\ -0.5028 & 0.6781 \\ -7.9532 & 7.3762 \end{bmatrix}, \tag{42}$$

$$C = \begin{bmatrix} 0.4723 & 0.8878 & 0.1904 & 0.5784 \\ 0.5252 & 0.4876 & 0.6927 & 0.0413 \end{bmatrix}. \tag{43}$$

The input constraints are assumed to be

$$-2 \leq \tilde{u}_1 \leq 2 \tag{44}$$

$$-2 \leq \tilde{u}_2 \leq 2 \tag{45}$$

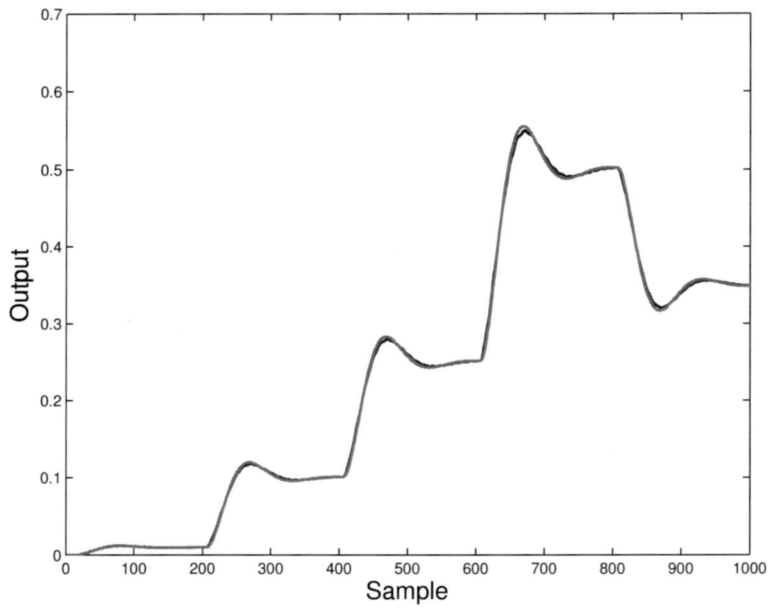

Figure 5. Comparison between both robust control algorithms. Blue: performance achieved with the algorithm presented in this paper. Magenta: and the output obtained with the algorithm proposed by Bloemen et al. {2001}.

The algorithm developed in Section 3 is used to achieve a robust MPC control of this MIMO Hammerstein model. The following design parameters are considered: $Q_x = [1\ \ 0; 0\ \ 1]$; $R_u = [10\ \ 0; 0\ \ 5]$.

To calculate the sequence of control inputs, the RMPC algorithm requires the uncertain model which represents the observed behavior of the process. For this purpose, the linear block uncertainty was modeled as two possible values for A ($A_1(1,1) = 0.3780$ and $A_2(1,1) = 0.6780$). As regards the nonlinear block uncertainty, the following perturbations were considered: $\tilde{v}_1 = \tilde{u}_1 + 0.22\tilde{u}_1\tilde{u}_2 - 0.08\tilde{u}_1\tilde{u}_2^2 + 0.98\tilde{u}_2$ and $\tilde{v}_2 = -\tilde{u}_1 - 0.25\tilde{u}_1\tilde{u}_2 + 0.45\tilde{u}_1^2\tilde{u}_2 + 1.9\tilde{u}_2$. The polytopic representations for v_1 and v_2 are shown in Figures (6-7).

The setpoint sequence for the first output (y_1) is $0, 1, -0.5, 0$, while for the second output (y_2) is $0, 1.25, -0.75$.

In Figures (8-9) the simulations are plotted for the robust MPC algorithm herein presented.

As the optimization problem described in Theorem 3.1 should be solved at each sample time, it is interesting to consider the necessary time to obtain the solution. The implementation performed in this work made use of MATLAB® LMI Toolbox (Gahinet et al. {1995}). Specifically, the "mincx" routine was used for minimizing a linear objective under LMI constraints. The problem was formulated off-line, and only the matrix described in Eq. (26) was adapted at each sample time. In the complete simulation, accomplished with an Intel Processor Core I3 330M (2.13 GHz), the largest time required to perform the optimization was 1.2 seconds. This time demand results more than appropriate for most of

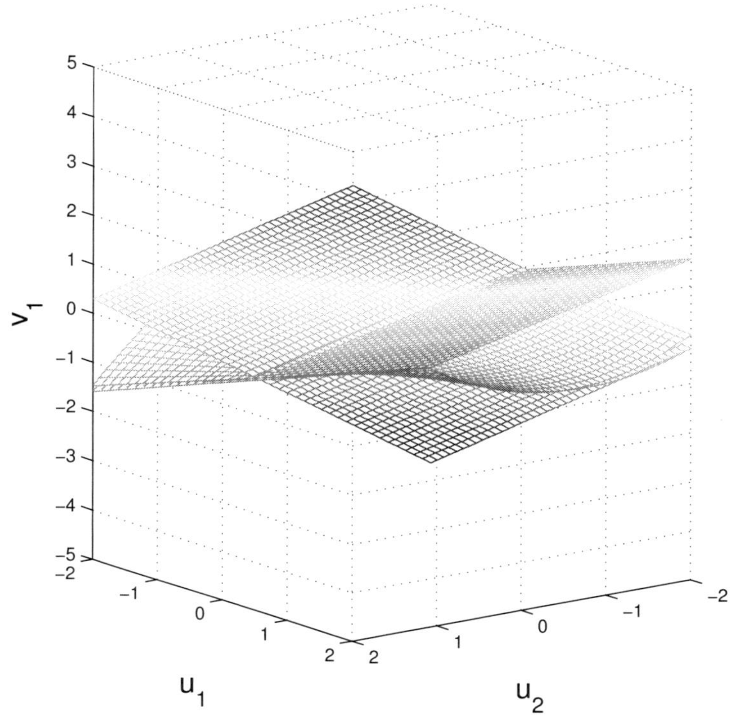

Figure 6. The static input nonlinearity: nominal value for $v_1(u_1, u_2)$, and polytopic description for this variable (i.e. the region between both planes).

the process applications.

To evaluate the controller performance under different conditions, four different situations are considered:

- Case I: nominal conditions for both the nonlinearity and the linear dynamics (i.e. Eqs. 38-39 and 41-43).
- Case II: nominal nonlinearity and uncertain linear dynamics.
- Case III: nominal linear block and uncertain nonlinearity.
- Case IV: uncertainty in both blocks (the linear dynamics uncertainty considered in case II and the uncertain nonlinearity in case III).

The required control movements predicted by the robust MPC algorithm for each of the four situations are depicted in Figures (10-11).

5. Discussion

The proposed controller design method addresses the problem of robust synthesis. Note that if the model predictive controller is designed without an explicit consideration of the

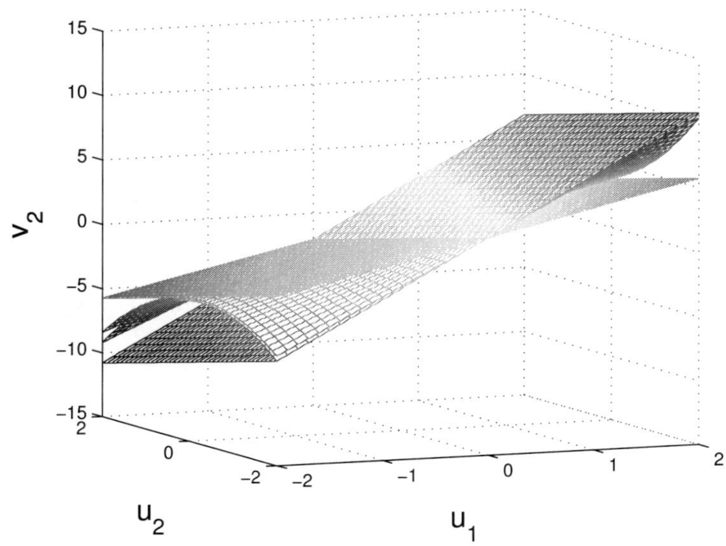

Figure 7. The static input nonlinearity: nominal value for $v_2(u_1, u_2)$, and polytopic description for this variable (i.e. the region between both planes).

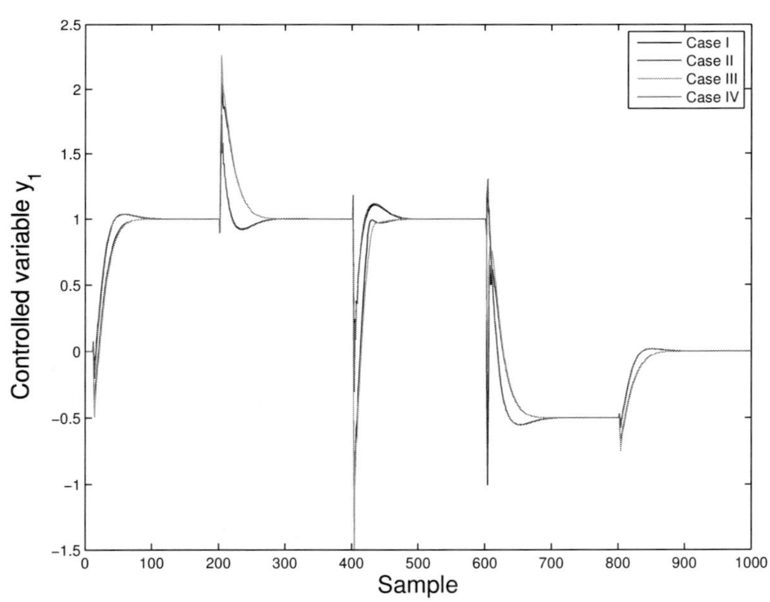

Figure 8. Output y_1: Performance achieved with the proposed robust MPC controller for the MIMO Hammerstein model. Four different situations were considered: nominal conditions for both the nonlinearity and the linear dynamics (Case I); nominal nonlinearity and uncertain linear dynamics (Case II); nominal linear block and uncertain nonlinearity (Case III); uncertainty in both blocks (Case IV).

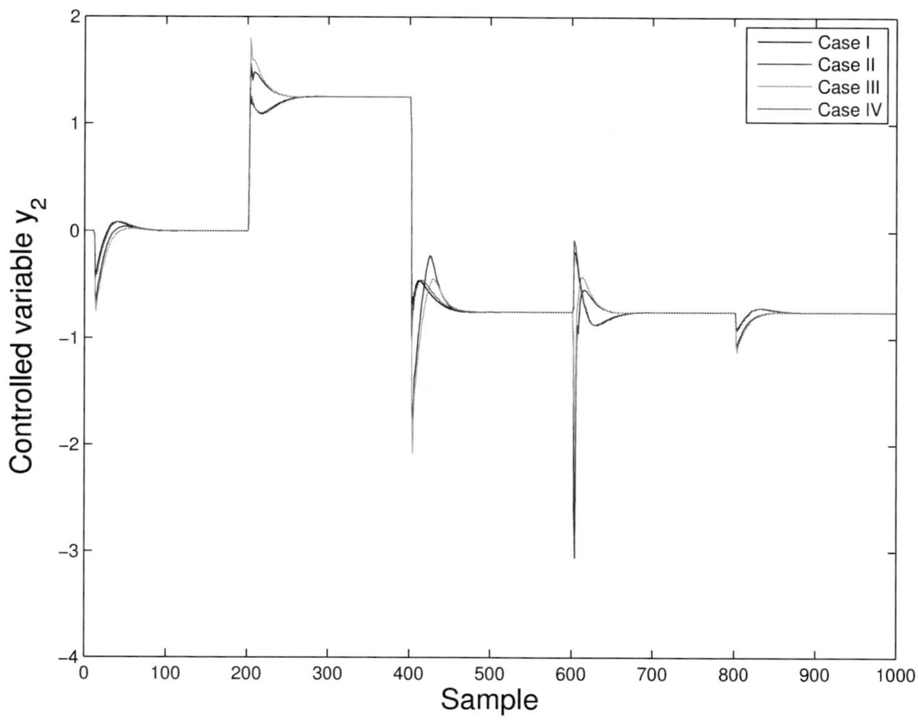

Figure 9. Output y_2: Performance achieved with the proposed robust MPC controller for the MIMO Hammerstein model. Four different situations were considered: nominal conditions for both the nonlinearity and the linear dynamics (Case I); nominal nonlinearity and uncertain linear dynamics (Case II); nominal linear block and uncertain nonlinearity (Case III); uncertainty in both blocks (Case IV).

plant uncertainty, the controlled system output could be unstable.

The unfavorable situation in which uncertainty can be present in both the linear and the nonlinear blocks is taken into account. From this point of view, the approach followed in this work can be interpreted as an extension of the algorithm in Kothare et al. {1996} for uncertain linear systems to Hammerstein structures with uncertainty.

At each control interval, the proposed robust MPC algorithm computes an open-loop sequence of manipulated variable adjustments in order to optimise the uncertain Hammerstein system behaviour. The first input in the optimal sequence is injected into the plant, and the entire optimisation is repeated at subsequent control intervals. The unconstrained algorithm includes $2L + 1$ LMIs, where L is the number of constitutive elements of the convex hull considered for the polytopic description of the linear dynamics. The other LMI is related to the required condition of achieving a stable feedback control loop. As a consequence of the reduced number of LMIs, the computational time demand is quite lower than the one involved in the approach in Bloemen et al. {2001}. Another difference with the mentioned work is that the approach herein considered uses a unique objective function, while the procedure in Bloemen et al. {2001} requires the performance index to be split into two parts (Chen and Allgöwer {1998}) for determining the manipulated variable.

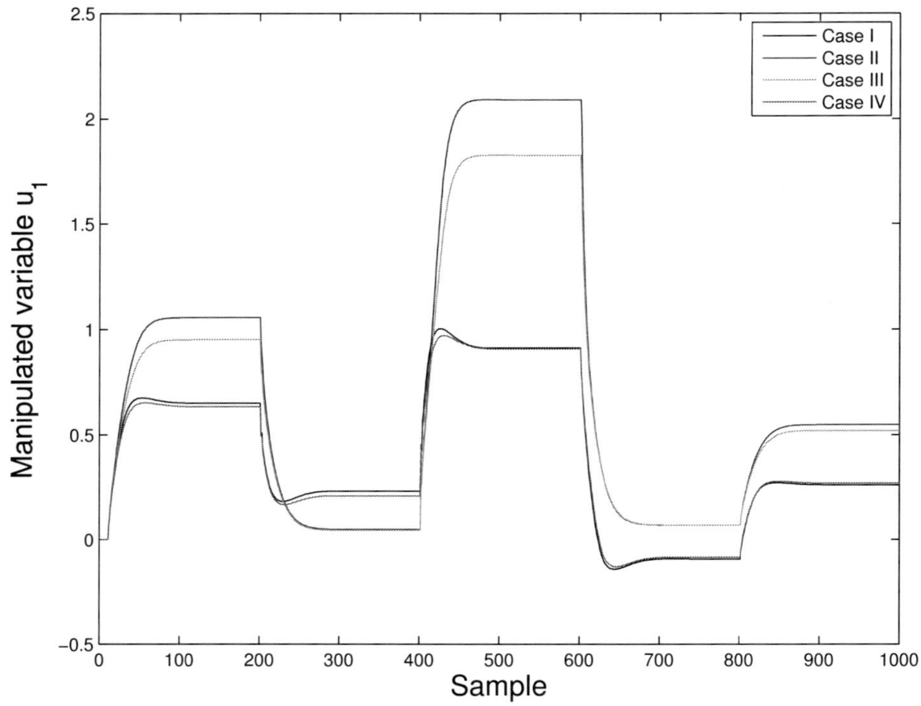

Figure 10. Calculated movements for the manipulated variable u_1. Four different situations were considered: nominal conditions for both the nonlinearity and the linear dynamics (Case I); nominal nonlinearity and uncertain linear dynamics (Case II); nominal linear block and uncertain nonlinearity (Case III); uncertainty in both blocks (Case IV).

As shown in Figure 2, the design criterion does not require inversion of the static nonlinearity. Then, the advantage is twofold: i) the design method is viable even when the nonlinearity inverse does not exist (for instance in those cases in which the nonlinearity describes a saturation behavior) and, ii) the design can be accomplished by considering the system behavior from manipulated variable to controlled variable and including this knowledge in the algorithm synthesis.

6. Conclusion

The main contribution of the present chapter relies in the development of an ad hoc robust model predictive based methodology for controlling both SISO and MIMO Hammerstein systems in the presence of uncertainty. It is assumed that the uncertain system is modeled following the polytopic approach both for the nonlinear and the linear blocks.

Representing the static nonlinearity uncertainty as polytopic descriptions, enables one to use a robust linear MPC algorithm. In this way, the problem remains as a convex optimization one.

The whole control problem (which deals with model uncertainty as well as input and output constraints) is formulated in the framework of LMI theory. For this purpose, Lya-

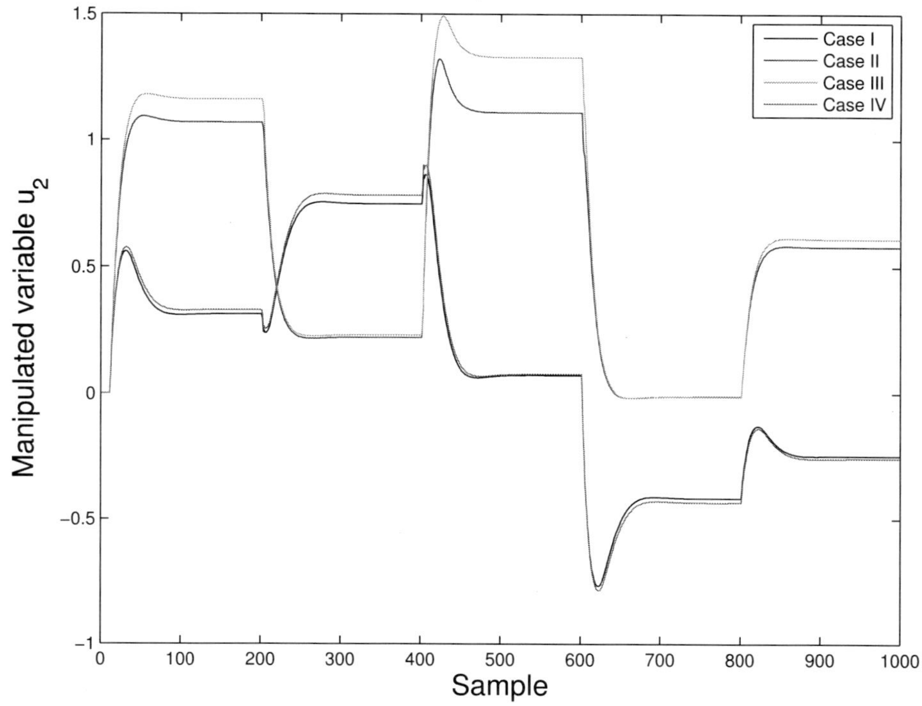

Figure 11. Calculated movements for the manipulated variable u_2. Four different situations were considered: nominal conditions for both the nonlinearity and the linear dynamics (Case I); nominal nonlinearity and uncertain linear dynamics (Case II); nominal linear block and uncertain nonlinearity (Case III); uncertainty in both blocks (Case IV).

punov functions are used. Therefore, an LMI optimization problem is stated and solved. The algorithm, which minimizes a "worst-case" infinite horizon objective function, incorporates input and output constraints and a description of the plant uncertainty and guarantees certain robustness properties.

The designed controller shows an appropriate performance in the tracking of changing setpoints. This results in an efficient robust control of uncertain Hammerstein systems.

Acknowledgments

This work was financially supported by the ANPCyT, CONICET and Universidad Nacional del Sur.

Appendix

Recalling Eqs. 2-3:

$$\begin{aligned} \tilde{x}(k+1) &= \tilde{A}\tilde{x}(k) + \tilde{B}\tilde{v}(k), \\ \tilde{y}(k) &= \tilde{C}\tilde{x}(k), \end{aligned}$$

and rewriting Eq. 2, but considering a step back in time, we obtain:

$$\tilde{x}(k) \;=\; \tilde{A}\tilde{x}(k-1) + \tilde{B}\tilde{v}(k-1). \tag{46}$$

Subtracting Eq. 46 from Eq. 2 gives:

$$\tilde{x}(k+1) - \tilde{x}(k) \;=\; \tilde{A}\left[\tilde{x}(k) - \tilde{x}(k-1)\right] + \tilde{B}\left[\tilde{v}(k) - \tilde{v}(k-1)\right], \tag{47}$$

and introducing the delay operator q^{-1} we obtain:

$$\tilde{x}(k+1) - \tilde{x}(k+1)q^{-1} \;=\; \tilde{A}\left[\tilde{x}(k) - \tilde{x}(k)q^{-1}\right] + \tilde{B}\left[\tilde{v}(k) - \tilde{v}(k)q^{-1}\right], \tag{48}$$

$$(1-q^{-1})\tilde{x}(k+1) \;=\; \tilde{A}(1-q^{-1})\tilde{x}(k) + \tilde{B}(1-q^{-1})\tilde{v}(k), \tag{49}$$

$$\Delta\tilde{x}(k+1) \;=\; \tilde{A}\Delta\tilde{x}(k) + \tilde{B}\Delta\tilde{v}(k). \tag{50}$$

Now, rewriting Eq. 3, but considering a step forward in time, we obtain:

$$\tilde{y}(k+1) \;=\; \tilde{C}\tilde{x}(k+1), \tag{51}$$

and subtracting equation Eq. 3 from Eq. 51 gives:

$$\tilde{y}(k+1) - \tilde{y}(k) \;=\; \tilde{C}(1-q^{-1})\tilde{x}(k+1). \tag{52}$$

Substracting the setpoint y_{sp} from both sides in Eq. 52, and replacing $x(k+1)$ by Eq. 2, gives the following expressions:

$$\tilde{y}(k+1) - \tilde{y}(k) - y_{sp} \;=\; \tilde{C}(1-q^{-1})[\tilde{A}\tilde{x}(k) + \tilde{B}\tilde{v}(k)] - y_{sp}, \tag{53}$$

$$\tilde{y}(k+1) - y_{sp} \;=\; \tilde{C}\tilde{A}\Delta\tilde{x}(k) + \tilde{C}\tilde{B}\Delta\tilde{v}(k) + \tilde{y}(k) - y_{sp}, \tag{54}$$

Note that Eqs. (50) and (54) can be rewritten in a matrix form, obtaining Eq. (4).

References

Alessio, A., Bemporad, A., 2009. A survey on explicit model predictive control. In: D.M. Raimondo L. Magni, F. A. (Ed.), *Lecture Notes in Control and Information Science*. Vol. 384. Springer-Verlag, Berlin, Ch. Nonlinear Model Predictive Control: Towards New Challenging Applications, pp. 345–369.

Araújo, H., Oliveira, G., 2009. An LMI approach for output feedback robust predictive control using orthonormal basis functions. In: *Proc. Joint 48th IEEE Conference on Decision and Control and 28th Chinese Control Conference*. December 16-18, Shangai, P.R. China.

Bai, E., Li, K., 2010. Convergence of an iterative algorithm for a general Hammerstein system identification. *Automatica* 46, 1891–1896.

Biagiola, S., Agamennoni, O., Figueroa, J., 2004. H_∞ control of a Wiener type system. *International Journal of Control* 77, 6, 572–583.

Biagiola, S., Figueroa, J., 2009. Wiener and Hammerstein uncertain models identification. *Mathematics and Computers in Simulation* 11, 79, 3296–3313.

Biagiola, S., Figueroa, J., 2011a. Identification of uncertain MIMO Wiener and Hammerstein models. *Computers and Chemical Engineering* 35, 12, 2867–2875.

Biagiola, S., Figueroa, J., 2011b. Robust model predictive control of Wiener systems. *International Journal of Control* 84, 3, 432–444.

Bloemen, H., Boom, T., Verbruggen, H., 2001. Model-based predictive control for Hammerstein-Wiener systems. *International Journal of Control* 74, 482–495.

Boyd, S., Ghaoui, L. E., Feron, E., Balakrishnan, V., 1994. *Linear Matrix Inequalities in System and Control Theory*. SIAM, Philadelphia, USA.

Camacho, E., Alba, C. B., 2004. Model Predictive Control. In: *Advanced Textbooks in Control and Signal Processing*, Springer-Verlag, 2nd ed., London, U.K.

Chan, K., Bao, J., 2007. Model predictive control of Hammerstein systems with multivariable nonlinearities. *Ind. Eng. Chem. Res.* 46, 168–180.

Chen, H., Allgöwer, F., 1998. A quasi-infinite horizon nonlinear model predictive control scheme with guaranteed stability. *Automatica* 34, 1205–1217.

Chen, J., Wang, X., Ding, R., 2012. Gradient based estimation algorithm for Hammerstein systems with saturation and dead-zone nonlinearities. *Applied Mathematical Modelling* 36, 238–243.

Dempsey, E., Westwick, D., 2004. Identification of Hammerstein models with cubic spline nonlinearities. *IEEE Transactions on Biomedical Engineering* 51, 2, 237–245.

Dolanc, G., Strmčnik, S., 2008. Design of a nonlinear controller based on a piecewise-linear Hammerstein model. *Systems and Control Letters* 57, 332–339.

Durfee, W., Maclean, K., 1989. Methods for estimating the isometric recruitment curve of electrically stimulated muscle. *IEEE Transactions on Biomedical Engineering* 36, 654–667.

Eskinat, E., Johnson, S., Luyben, W., 1991. Use of Hammerstein models in identification of nonlinear systems. *AIChE Journal* 37, 255–268.

Fruzzetti, K., Palazoglu, A., McDonald, K., 1997. Nonlinear model predictive control using Hammerstein models. *Journal of Process Control* 7, 31–41.

Gahinet, P., Nemirovski, A., Laub, A., Chilali, M., 1995. *LMI Control Toolbox*. The MathWorks, MA, USA.

Gómez, J., Baeyens, E., 2004. Identification of block-oriented nonlinear systems using orthonormal bases. *Journal of Process Control* 14, 685–697.

Hunter, I., Korenberg, M., 1986. The identification of nonlinear biological systems: Wiener and Hammerstein cascade models. *Biological Cybernetics* 55, 135–144.

Janczak, A., 2005. *Identification of Nonlinear Systems Using Neural Networks and Polynomal Models.* Springer, Berlin.

Kothare, M., Balakrishnan, V., Morari, M., 1996. Robust constrained model predictive control using linear matrix inequalities. *Automatica* 32, 1361–1379.

Le, F., I. Markovsky, C. F., Rogers, E., 2010. Identification of electrically stimulated muscle models of stroke patients. *Control Engineering Practice* 18, 396–407.

Mattsson, P., Wigren, T., 2014. Recursive identification of hammerstein models. In: *Proc. of American Control Conference*. Vol. June 4-6. Portland, Oregon, USA, pp. 2498–2503.

Mayne, D., Rawlings, J., Rao, C., Scokaert, P., 2000. Constrained model predictive control: Stability and optimality. *Automatica* 36, 789–814.

Norquay, S., Palazoglu, A., Romagnoli, J., 1998. Model predictive control based on Wiener models. *Chemical Engineering Science* 53, 1, 75–84.

Patwardhan, R., Lakshminarayanan, S., Shah, S., 2003. Constrained nonlinear MPC using Hammerstein and Wiener models: PLS framework. *AIChE Journal* 44, 7, 1611–1622.

Pearson, R., Pottmann, M., 2000. Gray-box identification of block-oriented nonlinear models. *Journal of Process Control* 10, 301–315.

Pomerleau, D., Hodouin, D., Poulin, E., 2003. Performance analysis of a dynamic phenomenological controller for a pellet cooling process. *Journal of Process Control* 13, 139–153.

Sentoni, G., Agamennoni, O., Desages, A., Romagnoli, J., 1996. Approximate models for nonlinear process control. *AIChE Journal* 42, 2240–2250.

Sjöberg, J., Zhang, O., Ljung, L., Benveniste, A., Delyon, B., Glorennec, P., Hjalmarsson, H., Juditsky, A., 1995. Nonlinear black-box modeling in system identification: A unified overview. *Automatica* 12, 12, 1691–1724.

Veselý, V., Rosinová, D., Foltin, M., 2010. Robust model predictive control design with input constraints. *ISA Transactions* 49, 114–120.

Westwick, D., Kearney, R., 2001. Separable least squares identification of nonlinear Hammerstein models: Application to stretch reflex dynamics. *Ann. Biomed. Eng.* 29, 8, 707–718.

In: Model Predictive Control
Editor: Corrine Wade, pp. 99-115
ISBN: 978-1-63463-859-3
© 2015 Nova Science Publishers, Inc.

Chapter 5

MODEL PREDICTIVE CONTROL IN HILBERT SPACE

A. I. Propoi*
Institute of Systems Analysis,
Russian Academy of Sciences, Moscow, Russia

Abstract

In model predictive control (MPC) the basic notions are a trajectory and a set of feasible trajectories. By using these notions we describe the MPC procedure in Hilbert space. It is shown that a feasible set of trajectories can be associated with a bounded positive self-adjoint linear operator in Hilbert space. This operator gives a complete description of CS (at a given time or a state). In the same way a target of control is described. We show that the principal stage of MPC — reconstruction of a new trajectory — can be formalized as the Newton transformation of a given trajectory. We show how these constructions in Hilbert space can be transformed into a mathematical (identified) model. For this an operator defined by relations between two bases (in Hilbert and finite coordinate spaces) is introduced and called a realization operator. This approach is generalized on a collection of trajectories (in Hilbert and coordinate spaces respectively) and naturally leads to frame theory. A predictive frame that gives a complete description of CS (at a given time or a state) is introduced in Hilbert space. By using the realization operator we transform the predictive frame into a table of numbers and we call it data predictive matrix. The data predictive matrix is a basis of a new approach to predictive control which we call data predictive control (DPC). We analyze the stages of DPC and show how by using different control strategies the construction of data predictive matrix can be adopted to goals of control. Finally, we consider a construction of the feasible trajectory which is the closest to a target (but unfeasible) trajectory (this procedure is similar to the consistent reconstruction in frame theory).

Keywords: model predictive control, feasible set, frame theory

AMS Subject Classification: 93-02, 92C10, 93D15, 42C15, 42A38

*E-mail address: propoi@gmail.com

1. Introduction

Model predictive control (MPC) is now a basic tool for implementation of control in a closed-loop mode for applications that require accounting different constraints on variables [1], [2], [3].

The idea of MPC suggested in [4] is simple and follows a common practice in decision making. It can be described in the following terms.

A control system (CS) at a current time generates a set of feasible trajectories. The feasible trajectories are predictive by their nature: each trajectory predicts the future behavior of CS for a fixed sequence of control actions during a given period of time (the planning horizon).

At this stage the trajectories are not fixed, they are only limited by dynamic and static constraints within the chosen mathematical model of CS. At the next stage an optimal trajectory is selected from this set that optimizes a given performance index and is implemented for some period of time (smaller than the planning horizon). At the end of this time period the current state is measured and the procedure is repeated for this new initial state.

Therefore, the basic objects in MPC are a trajectory and a feasible set of trajectories.

Considering an object as an element of Hilbert space is a common approach in many research areas such as quantum mechanics, signal processing, image analysis, information retrieval, etc. Following this practice we describe the MPC procedure in an abstract (real) Hilbert space.

A feasible subspace spanned by trajectories of a given CS is introduced. This subspace is characterized by a metric associated with a bounded positive self-adjoint linear operator. Using this metric we define the feasible set as a ball in this metric. Thus, this operator gives a complete description of CS (at a given time or a state).

In a similar way a target subspace with the corresponding metric is introduced.

In quantum mechanics such kind of operators is called observables [5]. They describe properties of a system. We note that quantum mechanics is probabilistic by its nature. Therefore a feasibile set of CS in general case should be also interpreted in probabilistic terms (i.e., as a concentration ellipsoid) [6].

The principal stage of MPC is the recalculation of optimal control for the current state, or, in other words, the reconstruction of the current trajectory. We show that in terms of the introduced operators (i.e., in the feasible and target subspaces) this stage can be formalized as the Newton transformation of a given trajectory.

In general case mathematical models of CS are nonlinear and contain a large number of constraints.

In this chapter we suggest a new approach to predictive control based on frame theory [7],[8] and called data predictive control (DPC) [6]. (The importance of the frame theory in control systems was underlined in [9].)

The basic features of DPC are following.

Instead of a mathematical model of CS a collection of trajectories generated by CS (called predictive frame) is used.

The trajectories of this frame span the feasible subspace and define a metric of the subspace. Therefore, the predictive frame gives a complete characterization of CS at a given state.

It is important to underline that both linear and nonlinear CS can be treated in the same way in this framework. Besides, trajectories of this frame are feasible by construction.

In frame theory the basic operations are measurement and reconstruction. We show how by using a measurement operator an abstract trajectory can be transformed into a sequence of numbers. Consequently, the frame in Hilbert space is transformed into a table of numbers. We call this table data predictive matrix.

The reconstruction of the current trajectory, that is, the construction of an optimal trajectory for a new state, is performed by a reconstruction operator which is reduced to computing a weighted sum of trajectories in the frame. The weight of a trajectory is considered as a control variable and is defined as a value of a linear functional for this trajectory. The linear functional describes a property of a trajectory. We show that this linear functional should be dual to the given trajectory (either in the metric of a feasible subspace or in the metric of a target subspace).

We analyze the stages of DPC and show how by using different control strategies the construction of data predictive matrix can be adopted to goals of control. Finally, we consider a construction of the feasible trajectory which is the closest to a target (but unfeasible) trajectory (this procedure is similar to consistent reconstruction in frame theory).

2. Basic Notions

In MPC, basic notions are a trajectory and a set of feasible trajectories [12].

If ψ is a trajectory, then $u\psi$, u is a real number, is also a trajectory (u is the length or the duration of the trajectory $u\psi$ along ψ), if ψ_1 and ψ_2 are trajectories, then $\psi_1 + \psi_2$ is a trajectory too (it is a path along ψ_1 and then along ψ_2). So, the space of trajectories is a linear space L.

We distinguish a real trajectory that has been realized by CS and a possible trajectory that can be realized by CS. The realization of a trajectory means the choice and the implementation of a sequence of control actions. Therefore we will also call a real trajectory the trajectory that has been computed from a mathematical model of CS for a given control.

The properties of a real trajectory can be measured (or estimated). On the other hand, a possible trajectory cannot be measured. However the set of possible trajectories must obey some constraints determined by the characteristics of a given CS. These constraints should be reflected in a chosen mathematical model of CS.

The most general type of such constraints is belonging of trajectories to some subspace of the trajectory space. In other words, CS at a given time t generates trajectories that span a subspace $S \subseteq L$.

Thus each trajectory can be associated with an one-dimensional subspace of the trajectory space L, and the set of possible trajectories (at a given time) — with a subspace $S \subseteq L$.

Borrowing basic notions from quantum mechanics [5] we call one-dimensional subspace of the trajectory space a pure state of a CS, correspondingly, a subspace of the trajectory space defines a general state of a CS. (We will also call it the feasible subspace of a CS).

The next type of the specification of a trajectory is its length or norm (e.g., determined by the duration of a trajectory). In this case we can associate with the given trajectory the vector $\psi \in L$ and with trajectories of a fixed duration that CS can generate at a given time t — the feasible set D_t of the feasible subspace S_t. Thus the trajectory space becomes a normed space and, with some additional requirements, a Banach space.

We shall assume that the feasible set D_t is a compact set in the feasible subspace S_t. If, in addition, this set is symmetric: $\psi \in D_t \Leftrightarrow -\psi \in D_t$, then the feasible set defines a norm $\|\cdot\|_t = \|\cdot\|_{D_t}$. In this norm the feasible set becomes a ball in the feasible subspace.

We introduce a continuous (bounded) linear functional φ on the trajectory space L and consider it as a property of a trajectory. The set of these functionals forms the dual space L^*; a bilinear form on the $L \times L^*$ defines the estimation of the property φ for the trajectory ψ.

For a given trajectory ψ it is important to determine the proper (optimal) property φ_ψ and, vice versa, for a given property φ — the proper (optimal) trajectory ψ_φ. It can be done by the duality relations.

In Banach space L we define the duality mapping $L \to L^*$:

$$T(\psi) = \{\varphi \in L^* \mid (\varphi, \psi) = \|\varphi\|\|\psi\|,\ \|\varphi\| = \|\psi\|\}. \tag{1}$$

The duality mapping was introduced in [10] and developed in many papers (see, e.g. [11]).

In general case the duality mapping is multivalued. We will assume that (1) defines a bijection: to each $\psi \in L$ corresponds the unique $\varphi_\psi \in L^*$ and to each $\varphi \in L^*$ corresponds the unique $\psi_\varphi \in L$. We shall relate the duality mapping (1) to the subspace S_t, i.e. to the norm $\|\cdot\|_t$.

If the feasible set satisfies some symmetry requirements then this set becomes an ellipsoid; or, in other words, if the norm of the trajectory space satisfies some conditions (such as the parallelogram law), then this norm induces an inner product. With some additional requirements the space with an inner product becomes a Hilbert space.

From now on we will assume that the trajectory space is a real Hilbert space H.

The set of trajectory properties determines the dual space H^*. The Riesz representation theorem establishes a connection between dual spaces H and H^* and, correspondingly, between bilinear forms on $H \times H^*$ and $H \times H$ (for real space it is the isometric isomorphism). Thus, each trajectory can be considered as a property and each property as a trajectory. Correspondingly, a subspace of trajectories can be considered as a subspace of properties and vice versa.

In more details the feasible set is analyzed in [12].

In the next Section we will show how, by using these basic notions we can describe the model predictive control in an abstract Hilbert space.

3. MPC in Hilbert Space

We begin with the conventional description of model predictive control [2].

Let CS be described by the system of differential equations

$$\dot{x}(t) = f(x(t), u(t)),\ x(0) = x_0,\ t \in [0, T], \tag{2}$$

subject to input and state constraints:

$$x(t) \in D_x,\ u(t) \in D_u. \tag{3}$$

Here $x(t) \in R^n$ and $u(t) \in R^m$ are the (phase) state and control (inputs) variables, correspondingly.

The perfomance index is

$$J = \int_0^T f_0(x(t), u(t))dt \longrightarrow \max. \tag{4}$$

By considering the perfomance index in the differential form

$$\dot{x}_0(t) = f_0(x(t), u(t)),\ x_0(0) = 0, \tag{5}$$

and introducing the extended phase space R^{n+1} with variables $\{x_0(t), x_1(t), \ldots, x_n(t)\}$ as an extended phase state $\tilde{x}(t)$, the problem (1) – (4) can be rewritten in the form

$$d\tilde{x}(t)/dt = \tilde{f}(\tilde{x}(t), u(t)),\ \tilde{x}(0) = (0; x_0),\ t \in [0, T], \tag{6}$$

$$J = (\tilde{e}, \tilde{x}(T)) \longrightarrow \max,\ \tilde{e} = (1; 0), \tag{7}$$

subject to input and state constraints.

The problem (6), (7) can be generalized for an arbitrary linear functional $J = (\tilde{c}, \tilde{x}(T))$, where $\tilde{c}$ is the given vector in R^{n+1}, and for nonlinear functional $J = \Phi(\tilde{x}(T))$. Further we will not distinguish the phase state $x(t) \in R^n$ and its extended version.

In the phase state R^n we define the feasibility state $R_T(x_0)$. It is the set of all states $x(T)$ which can be achieved by CS (2), (3) using feasible controls for the time period (planning horizon) $[0, T]$. Then the optimization problem for the initial state x_0 can be stated as

$$\max_{x \in R_T(x_0)} \Phi(x). \tag{8}$$

Let $\bar{x} = \bar{x}(T)$ be a solution of (8). We define the vector $\bar{v} = \bar{x} - x_0$ in R^n from the initial state x_0 to the optimal terminal state $\bar{x}$. If the solution of the optimal control problem was obtained by using the differential equation (2), then we get the optimal trajectory (a curve in R^n) $\bar{x}(t), t \in [0, T]$, starting at x_0 and ending at $\bar{x}(T)$.

In MPC, the optimal trajectory $\bar{x}_{[0,T]}$ is used for some period of time $[0, t_1]$, $0 < t_1 < T$, then a new state $x_1 = x(t_1)$ is measured and the optimal control is recalculated for this new initial state x_1.

We will now describe the MPC in Hilbert space.

We consider the space of trajectories as a real Hilbert space H; ψ, $(\psi, \psi) = 1$, is a trajectory of duration $[0, T]$. For the current time t we define $S_t \subset H$ — the feasible subspace of CS, $D_t \subset S_t$ — the feasible set of CS.

We assume that:

1) D_t is a compact in the subspace S_t (closed bounded set that contains interior points in S_t);

2) D_t is a convex set: if trajectories ψ_1, ψ_2 are feasible, then the trajectory $\xi_1\psi_1 + \xi_2\psi_2$, $\xi_1, \xi_2 \geq 0$, $\xi_1 + \xi_2 = 1$, is also feasible;

3) D_t is a central symmetric body in S_t: $\psi \in D_t \Leftrightarrow -\psi \in D_t$;

4) each hyperplane section of D_t has a center.

This means that D_t is an ellipsoid in the subspace S_t and therefore can be described as

$$D_t = \{\psi \mid (\psi, G_t\psi) \leq 1\}, \tag{9}$$

where G_t is a bounded positive self-adjoint ($G = G^*$) linear operator.

In quantum mechanics such operators are called observables; $(\psi, G\psi)$ is the average of observable G at the state ψ. A positive self-adjoint operator is frequently associated with energy in quantum mechanics [5].

Therefore, in general case the frontier of a feasible set should be understood in probabilistic terms (i.e., as a probability or as an average). In this case, (9) is a concentration ellipsoid.

Note also that frequently $S_t = S \equiv H$ for all t, then H is the trajectory space of a given CS.

The operator G_t induces a new metric in the subspace S_t: $\|\psi\|_G^2 = (\psi, G\psi)$.

The target of control can be defined in a similar way as a target subspace $\bar{S}_t$, or as a target set $\bar{D}_t \in \bar{S}_t$.

Let the frontier of the target set be described by a function $q_t(\psi)$. We again assume that $q_t(\psi)$ is a quadratic function: $q_t(\psi) = ((\psi_{t1} - \psi), Q_t(\psi_{t1} - \psi))$, where Q_t is a bounded positive self-adjoint linear operator, ψ_{01} denotes the center of the target set for $t = 0$.

Then MPC problem in the space H can be formulated as [6]

$$\sup_{\psi \in D_t} q_t(\psi). \tag{10}$$

The solution ψ_0 of this problem for $t = 0$ satisfies the equality

$$Q_0\psi_1 = G_0\psi_0, \tag{11}$$

where $\psi_1 = \psi_{01} - \psi_0$. In the equivalent form

$$\psi_1 = Q_0^{-1}G_0\psi_0. \tag{12}$$

Thus, CS, moving along the trajectory ψ_0 (being at the state ψ_0) at time $t = 0$ generates the set of feasible trajectories D_0 (and the target set $\bar{D}_0$). Then (12) defines the optimal change of states.

On the other hand, the equation

$$\psi_0 = G_0^{-1}Q_0\psi_1 \tag{13}$$

determines the optimal trajectory ψ_0, $\psi_1 = \psi_{01} - \psi_0$, for the given target $\bar{D}_0$ (with the center ψ_{01}).

Note that (12) is the Newton direction in optimization methods.

Note also that (13) is the gradient of the function q_t at ψ_1 in the metric, determined by G_t, $t = 0$.

The optimality conditions for the optimization problem are given by Hahn-Banach theorem. Namely, there exists the hyperplane defined by the linear functional φ_{01} that is the support plane both for feasible and target feasible sets:

$$\sup_{\psi \in D_0} (\varphi_{01}, \psi) = (\varphi_{01}, \psi_0), \tag{14}$$

$$\sup_{\psi \in \bar{D}_0} (\varphi_{01}, \psi) = (\varphi_{01}, \psi_1). \tag{15}$$

The norm of the subspace S_0 is defined as $\|\psi\|_0 = (\psi, G_0\psi)^{1/2}$. Then (14) becomes

$$\sup \frac{(\varphi, \psi)}{\|\psi\|_0} = (\varphi, \psi_0). \tag{16}$$

By definition, $(\varphi, \psi_0) = \|\varphi\|_0$, $\varphi \in H^*$, and we have

$$\sup \frac{(\varphi, \psi)}{\|\psi\|_0} = \|\varphi\|_0, \qquad \sup \frac{(\varphi, \psi)}{\|\varphi\|_0} = \|\psi\|_0. \tag{17}$$

Thus, the extremal problem (14) determines the duality mapping $H \to H^*$ (1) for the norm (16) defined by the linear operator G_0. By associating $\varphi^* \in H^*$ with $\varphi \in H$ the duality mapping can be written in the form of the transformation $G_0 : H \to H$

$$\varphi = G_0\psi \tag{18}$$

with $\|\psi\|_0 = (\psi, G_0\psi)^{1/2}$, $\|\varphi\|_0 = (\varphi, G_0^{-1}\varphi)^{1/2}$.

Similarly, the extremal problem (15) defines the duality mapping $H \to H^*$ (1) for the norm defined by the linear operator Q_0:

$$\varphi = Q_0\psi. \tag{19}$$

Thus, the motion of CS along the trajectory ψ_0 (at the state ψ_0) is characterized by the dual pair $\{\psi_0; \varphi_{01}\}$, where $\varphi_{01} = G_0\psi_0$.

Similarly, the motion of CS along the trajectory ψ_1 is characterized by the dual pair $\{\psi_1; \varphi_{01}\}$, $\varphi_{01} = Q_0\psi_0$ with the same linear functional φ_{01} but with the different metric Q_0.

We note that the target set $\bar{D}_0$ can be identified with the next feasible set D_1 (and the linear operator Q_0 with G_1).

Let us consider the motion along the trajectory ψ_0. This trajectory was determined at time $t = 0$ for the planning horizon $[0, T]$ and is optimal for the current time $t > 0$. The trajectory ψ_0 is related to the linear functional φ_{01} which is dual to ψ_0 in the metric G_0.

At time t_1, $0 < t_1 < T$, CS generates the new feasible set determined by the linear operator G_1.

If $G_1 = G_0$, then we have from (12)

$$\psi_0 = G_0^{-1}G_0\psi_0 \tag{20}$$

— CS continues its motion along the trajectory ψ_0.

If the feasible set at time t_1 differs from the feasible set at time t_0, then CS changes its trajectory

$$\psi_1 = G_1^{-1}G_0\psi_0. \tag{21}$$

Finally, we include target set (defined by Q_0) in these constructions. We assume that the target does not change for period $[0, t_1]$. Then

$$G_0\psi_0 = Q_0\psi,$$

$$G_1\psi_1 = Q_0\psi.$$

From these equations follows (21).

In this section a general MPC scheme was described in an abstract Hilbert space. This scheme uses a trajectory and a feasible set as basic notions. In the next sections we relate this abstract model to mathematical models in the coordinate space.

4. Predictive Frame

Let S be a feasible subspace. We consider a collection of trajectories $\psi_j \in S$, $j \in J$, and assume that any feasible trajectory $\psi \in S$ can be constructed as a linear combination of these trajectories:

$$\psi = \sum_{j\in J} \theta_j \psi_j. \tag{22}$$

Naturally, such a collection must satisfy some conditions.

Definition 1. *A countable collection of vectors $\psi_j \in S$, $j \in J$, is called a frame of a subspace S, if there exists positive constants A, B, such that*

$$A\|\psi\|^2 \leq \sum_{j\in J} (\psi_j \psi) \leq \|\psi\|^2. \tag{23}$$

The theory of frames for a Hilbert space plays a fundamental role in signal processing, image processing, data compression, sampling theory, etc [7], [8], [9].

We call the frame of the feasible subspace S the predictive frame of CS, it consists of trajectories that predict future behavior of CS (at the given state).

The equation (22) transforms a given sequence of numbers θ_j, $j \in J$, into the vector ψ in S. Denoting this sequence as θ and assuming that θ is an element of the space $l^2 = l^2(J)$ (square summable sequences of numbers) we can define operator $\Psi : l^2 \to S$

$$\Psi\theta = \psi. \tag{24}$$

In frame theory vectors $\psi_j \in S, j \in J$, are called reconstruction vectors, the subspace S — reconstruction subspace [7],[8]. We will call the feasible subspace $S = S_\Psi$ construction subspace, Ψ is construction operator.

In frame theory the rule of obtaining numbers θ_j is fixed and is called measurement. Namely, it is supposed that θ_j is the value of a linear functional $\varphi_j \in S^*$ for the vector ψ in S:

$$\theta_j = (\varphi_j, \psi). \tag{25}$$

The Hilbert space coincides with its dual: $H = H^*$. Hence, (24) can also be considered as the inner product of vectors φ_j, ψ in H.

The linear functional (vector) φ_j was interpreted as a property of a trajectory. Therefore, (25) describes the measurement (estimation) of a trajectory ψ with respect to a property φ_j.

The linear functionals φ_j, $\varphi_j \in H^*$, $j \in J$, span the subspace $S^*_\Phi \subset H^*$ ($S_\Phi \subset H$). This subspace defines a general property. In frame theory vectors φ_j are called measurement or sampling vectors, the subspace S_Φ — a measurement or sampling subspace. The operator $\Phi^* : S^* \to l^2$:

$$\Phi^*\psi = \theta \tag{26}$$

maps a given trajectory ψ to a sequence of numbers — the vector θ.

The operator Φ^* is called a measurement operator. The vector θ is a representation (or a model, or an image) of the trajectory ψ.

From (24) and (26) it follows that

$$\psi' = \Psi\Phi^*\psi, \tag{27}$$

or

$$\psi' = \sum_{j\in J} \psi_j \varphi_j^* \psi. \tag{28}$$

In the basic case, $S_\Psi = S_\Phi$ and $\Psi = \Phi$. Then

$$\psi' = \Psi\Psi^*\psi, \tag{29}$$

or

$$\psi' = \sum_{j\in J} \psi_j \psi_j^* \psi. \tag{30}$$

Thus, by using the combination of the two basic operations — measurement and construction we can determine the new trajectory ψ' for the given trajectory ψ.

As it will be shown later, the transformation (25), (26) describes not only the measurement of a given trajectory ψ but also the determination of control actions for the state ψ.

We denote

$$G = G_\Psi = \Psi\Psi^*. \tag{31}$$

The operator G is called the frame operator. It is shown that G is a bounded positive self-adjoint operator $S_\Psi \to S_\Psi$ [7],[8].

Hence, this operator is invertible: $G^{-1} = (\Psi\Psi^*)^{-1}$. (However, to invert $\Psi\Psi^*$ it is necessary to use pseudoinverse operators).

The operator G defines a metric on the subspace $S = S_\Psi$: $\|\psi\|_G^2 = (\psi, G\psi)$, $\psi \in S$, and, as a consequence, it determines the dual mapping $S \to S$, defined by the combination of the operator $\Psi^* : S \to l^2$ and $\Psi : l^2 \to S$.

From the above constructions follows that for any trajectory ψ:

$$\psi = G^{-1}G\psi = (\Psi\Psi^*)^{-1}(\Psi\Psi^*)\psi. \tag{32}$$

and

$$\psi = GG^{-1}\psi = (\Psi\Psi^*)(\Psi\Psi^*)^{-1}\psi. \tag{33}$$

Taking into account (30) and (32) we obtain

$$\psi = \sum_{j\in J} \tilde{\psi}_j \psi_j^* \psi = \tilde{\Psi}\Psi^*\psi, \tag{34}$$

where vectors ψ_j and $\tilde{\psi}_j$ are dual in the metric G:

$$\tilde{\psi}_j = G^{-1}\psi_j. \tag{35}$$

For (33), (35) we get

$$\psi = \sum_{j\in J} \psi_j \tilde{\psi}_j^* \psi = \Psi\tilde{\Psi}^*\psi. \tag{36}$$

Thus, we have the following scheme of controlling motion (cf (18) – (21)).

CS, moving along a given trajectory ψ_0, generates at time $t = 0$ a set of feasible trajectories (frame) $\Psi_0 = \{\psi_j\}$, $j \in J_0$.

By using this set the trajectory ψ_0 is estimated with respect to ψ_j: $(\psi_0, \psi_j) = \theta_j, j \in J$.

Considering the values θ_j, $j \in J$, as control actions and using the dual trajectories $\tilde{\psi}_j$, CS reconstructs the trajectory ψ_0, that is, continues its moving along ψ_0.

In (36) the dual vectors $\tilde{\psi}_j$ are used for the estimation of the trajectory ψ_0 and the vectors ψ_j — for the reconstruction of the trajectory ψ_0 .

In general case for the construction of a new trajectory a special frame $\Phi : \{\varphi_j\}$ should be introduced. (It depends on the goal of control). This frame defines the target subspace S_Φ with the metric $Q = \Phi\Phi^*$. This general scheme is described by relations (27) or (28). Note that in this scheme vectors φ_j in (28) are always dual to vectors ψ_j. However, they can be dual either in the matric $G_\Psi = \Psi\Psi^*$ and in this case CS continues its motion along ψ_0, or they are dual in the matric $G_\Phi = Q = \Phi\Phi^*$ and in this case CS changes its motion.

5. Representation of Trajectories

Let x_{it}, $i \in I(n), t \in I(T)$, be the value of the index i of the state (trajectory) ψ at the time t. We introduce the sequence of numbers (ordered by t) $\mathbf{x} = [x_{it}]$, $i \in I(n), t \in I(T)$ and call it the (original) representation of the trajectory ψ.

This sequence can be considered as a vector in the component space R^N, $N = nT$. In this case,

$$\mathbf{x} = \sum_{i\in I(n)} \sum_{t\in I(T)} x_{it}\mathbf{e}_{it}, \tag{37}$$

where $\mathbf{e}_{it} \in R^N$ are column vectors of zeros except for 1 in the position it.

These vectors form an orthonormal basis and therefore

$$x_{it} = (\mathbf{e}_{it}, \mathbf{x}), \tag{38}$$

where $(\cdot, \cdot)$ is the inner product in R^N.

By using (38), we can be rewrite (37) as

$$\mathbf{x} = \sum_{i\in I(n)} \sum_{t\in I(T)} \mathbf{e}_{it}\mathbf{e}_{it}^*\mathbf{x} = \sum_{i\in I(n)} \sum_{i\in I(T)} P_{it}\mathbf{x}, \tag{39}$$

where $\mathbf{e}_{\mathrm{it}}^*$ is a row vector (transpose to $\mathbf{e}_{\mathrm{it}}$), $P_{it} = \mathbf{e}_{\mathrm{it}}\mathbf{e}_{\mathrm{it}}^*$ is the one-dimensional projector.

We will call (37) – (39) the original representation of a trajectory $\psi \in H$ and the space $R^N = l^2(N)$ the model (representation) space of the trajectory space H.

To link this representation with the trajectory $\psi \in H$ we consider the following construction.

Let

$$\psi = \psi_S + \psi_S^\perp, \tag{40}$$

where $\psi \in H$, $\psi_S = P_S\psi$, P_S is the orthogonal projector onto subspace S, $\psi_S^{\perp} \in S^{\perp}$, $S^{\perp}$ is the orthogonal complement to S: $(\psi, \psi_S) = 0$, $\psi \in S$, $\psi_S^{\perp} \in S^{\perp}$.

We assume that the dimension of the subspace S equals N and choose in S an orthonormal basis $\varphi_{it}, it \in J(N)$. Then

$$P_S = \sum_{i \in I(n)} \sum_{t \in I(T)} \varphi_{it}\varphi_{it}^* = \sum_{i \in I(n)} \sum_{t \in I(T)} P_{it}. \tag{41}$$

In this case

$$\psi = P_S\psi = \sum_{i \in I(n)} \sum_{t \in I(T)} \varphi_{it}\varphi_{it}^*\psi = \sum_{i \in I(n)} \sum_{t \in I(T)} \theta_{it}\varphi_{it}, \tag{42}$$

where

$$(\varphi_{it}, \psi) = \theta_{it}. \tag{43}$$

By definition of projector $P_S\varphi_{it} = \varphi_{it}$. Therefore,

$$\theta_{it} = (\varphi_{it}, \psi) = (P_S\varphi_{it}, \psi) = (\varphi_{it}, P_S\psi). \tag{44}$$

We introduce the linear operators $C : S \to R^N$ and $C^* : R^N \to S$ and define them via relations between orthonormal bases φ_{it} in S and $\mathbf{e}_{it}$ in R^N:

$$C\varphi_k = \mathbf{e}_k, \tag{45}$$

$$C^*\mathbf{e}_k = \varphi_k, \tag{46}$$

where $k = \{it\} \in J(N)$, $J(N) = J(n) \cup J(T)$. From (45), (46) and the orthonormality bases follows that

$$C = \sum_{k \in J(N)} \mathbf{e}_k\varphi_k^*, \tag{47}$$

$$C^* = \sum_{k \in J(N)} \varphi_k\mathbf{e}_k^*. \tag{48}$$

Then (47), (48) give

$$CC^* = \sum_{k \in J(N)} \mathbf{e}_k\mathbf{e}_k^* = Id_N, \tag{49}$$

$$C^*C = \sum_{k \in J(N)} \varphi_k\varphi_k^* = P_S. \tag{50}$$

Definition 2. *A bounded linear operator C in Banach space is said to be a partial isometry if $\|C\psi\| = \|\psi\|$ for all $\psi \in Ker(C)^{\perp}$.*

From this definition and (47) – (48) follows that the operator C defined in (45), (46) is a partial isometry:

$$(\varphi, \psi)_S = (C^*\mathbf{e}, C^*\mathbf{x}) = (\mathbf{e}, CC^*\mathbf{x}) = (\mathbf{e}, \mathbf{x})_{R^N}. \tag{51}$$

Because of (45) – (50) the operator C is also called a transjector [13].

The operator C establishes the correspondence between an abstract model in H and a mathematical (identified) model in R^N.

Defining the operator $C = C_S$ by relations $C_S\varphi_{it} = \mathbf{e}_{it}$, where $\{\varphi_{it}\}$, $\{\mathbf{e}_{it}\}$ are orthonormal bases in $S \subset H$ and R^N, we obtain, instead of (39):

$$C_S\psi = \sum_{i\in I(n)}\sum_{t\in I(T)} C_{it}\psi = \mathbf{x} = \sum_{i\in I(n)}\sum_{t\in I(T)} x_{it}\mathbf{e}_{it}.$$

Evidently, (37) – (39) is not the unique representation of trajectory.

In particular, let

$$x_t = \sum_{i\in I(n)} C_{it}\psi = C_t\psi. \tag{52}$$

Then we get the representation x_t of the trajectory in the phase space R^n at time t.

Note that by using this representation we can index a trajectory ψ not by time $\psi_0 = \psi_{t_0}$, but by the current phase state $\psi_0 = \psi_{x_0}$.

Let

$$\mathbf{x}' = \sum_{k\in I(M)} c_k\mathbf{u}_k = \sum_{j\in I(n_1)}\sum_{\tau\in I(T_1)} c_{j\tau}u_{j\tau}, \tag{53}$$

where $\{\mathbf{u}_k\}$ is an orthonormal basis in R^M. Therefore $c_k = (\mathbf{u}_k, \mathbf{x})$ and consequently

$$\mathbf{x}' = \sum_{k\in I(M)} \mathbf{u}_k\mathbf{u}_k^*\mathbf{x} = P_u\mathbf{x}. \tag{54}$$

If we assume that $M = n_1T_1 = N = nT$, then for $T_1 < T$ should be $n_1 > n$ and vice versa. It shows an interrelation between realization of control actions at a given time and the distribution of control actions over time.

Thus, different representations of the trajectory space define different models of CS [6].

In particular, the feasible set defined by the quadratic form $(\psi, G\psi)$ in H is replaced by the quadratic form $(\mathbf{x}, G_x\mathbf{x})$ in R^N, where $C\psi = \mathbf{x}$ and

$$G = C^*G_xC, \tag{55}$$

the operator C is defined by the orthonormal bases in H and R^N (45), (46).

We call this operator a realization operator. It realizes a mathematical model of CS in the chosen set of indices i and t for the given abstract model of CS in H.

In general case (that is, not for orthonormal bases) a similar operator is defined in the frame theory (see the next section).

6. Data Predictive Control

We consider a collection of N_0 feasible trajectories $\psi(j) \in S$, $j \in J(N_0)$.

By choosing indices $i \in J(n)$ and time discretization $t \in J(T)$, we represent the trajectories $\psi(j) \in S$ by column-vectors $\mathbf{x}(j)$ that satisfy the equations (cf (2) – (7)):

$$x_{t+1}(j) = f(x_t(j), y_(j)),\ t = 0,\ldots,T-1,\ j \in I(N_0), \tag{56}$$

$$g(x_t(j), y_t(j)) \leq 0, \ x_0(j) = x, \ j \in I(N_0). \tag{57}$$

By measuring the outputs of CS (either from a model of CS or from real simulation) we obtain a table of numbers X. The columns of this matrix are trajectories $\mathbf{x}(j)$ of the dimension $N = nT$.

Each feasible control $y(j)$, $j \in J(N_0)$, produces the feasible trajectory $\mathbf{x}(j)$ in the space R^N. (We use here "trajectory" and "representation of a trajectory" as synonyms). Note, that by its nature the trajectory $\mathbf{x}(j)$ is predictive for the state $\mathbf{x}_0$.

Therefore instead of a mathematical model of CS we can deal with a data predictive matrix of the dimension $N \times N_0$ formed by the feasible trajectories $\mathbf{x}(j)$

$$X = [\mathbf{x}(1), \ldots, \mathbf{x}(N_0)]. \tag{58}$$

We assume that the trajectories $\mathbf{x}(j)$ in (58) span the feasible subspace S_x. Note, that $C : S \to S_x$, $S \subset H$, $S_x \subset R^N$.

Thus, the data matrix X give the complete description of CS within the chosen set of indices (at the given time). We call this matrix data predictive matrix.

The data predictive matrix (58) defines a (finite) frame in R^N: any feasible trajectory $\mathbf{x} \in S_x$ can be represented as

$$\mathbf{x} = \sum_{j \in J(N_0)} \mathbf{x}(j) z_j, \tag{59}$$

or

$$\mathbf{x} = X\mathbf{z}. \tag{60}$$

The data matrix X defines the mapping $R^{N_0} \to S_x$, $S_x \subset R^N$.

As an example we consider the linear CS described by equations

$$x_{t+1} = Ax_t + Bu_t, \, t = 0, 1, \ldots, T-1, \tag{61}$$

where $x_t \in R^n$, $u_t \in R^m$.

In the space of trajectories the equation (61) can be written as

$$\mathbf{x} = \mathbf{x}_0 + W\mathbf{u}, \tag{62}$$

where $\mathbf{x} = \{x_1, \ldots, x_T\}$, $\mathbf{x}_0 = \{x_0, \ldots, x_{T-1}\}$, $\mathbf{u} = \{u_1, \ldots, u_{T-1}\}$, W is the $N \times M$ matrix, defined by matrices A and B in (61) [6].

We will consider (62) in the form

$$\mathbf{x} = \mathbf{x}_0 + \sum_{k \in J(M)} \mathbf{w}_k y_k, \tag{63}$$

where the values y_k are defined by control variables u_{it}. (Note that $\mathbf{u} = \{u_{it}\} = \{y_k\} = \mathbf{y}$).

Formally, each vector $\mathbf{w}_k$ in R^N can be considered as a trajectory. An essential difference between these trajectories and the trajectory $\mathbf{x}_0$ is that a weighted sum of trajectories $\mathbf{w}_k$ can influence the trajectory $\mathbf{x}_0$ beginning only from the phase state x_1.

The matrix W determines mapping $R^M \to R^N$, where R^M is the space of controls $\mathbf{u}$. The image (range) of this map is given by

$$\mathrm{Im}(W) = \{\mathbf{x} \,|\, \mathbf{x} = W\mathbf{y}, \ \mathbf{y} \in R^M\}. \tag{64}$$

The feasible subspace at the state $\mathbf{x}_0$ is $S_0 = \mathbf{x}_0 + \mathrm{Im}(W)$. If CS (62) is controllable, then the matrix W has the full rank and the dimension of the feasible subspace equals to n.

Comparing (59) and (63) we see that W defines the (original) frame of CS. However, the representation of a feasible trajectory as a linear combination of trajectories $\mathbf{x}(j)$ in (59) provides more flexibility than the original model because one can adopt the data predictive matrix X to the specifics of the given optimal control problem.

For the linear system (62) let $Y = [\mathbf{y}(1), \ldots, \mathbf{y}(N_0)]$ be the $M \times N_0$ matrix of controls. Then (62) can be written in the matrix form

$$X = WY \tag{65}$$

(the trajectory $\mathbf{x}_0$ is omitted in (65)).

By using (65) we transform (60) into

$$\mathbf{x} = WY\mathbf{z} = W\mathbf{y}. \tag{66}$$

Therefore $Y\mathbf{z} = \mathbf{y}$ or

$$\sum_{j \in J(N_0)} \mathbf{y}(j) z_j = \mathbf{y}. \tag{67}$$

In control actions (67) becomes

$$\sum_{j \in J(N_0)} u_{it}(j) z_j = u_{it}, \tag{68}$$

for each $\{it\} \in J(N)$.

Generally, the transformation of the weights z_j into the real control actions u_{it} can be done only in the direction $\mathbf{z} \to \mathbf{u}$. The opposite direction can be fulfilled if the matrix Y is invertible, or, in general case, by using the pseudoinverse of Y, or the singular value decomposition (SVD) of Y.

We analyze different strategies of the generation of the data predictive matrix.

1) Let $N_0 = M$ and $Y_1 = I_{M \times M}$. Then (66) is reduced to (62): $X = W$, i.e., the data matrix X "digitizes" the original model W of CS.

2) Let

$$W = \hat{U}\hat{S}\hat{V}^* \tag{69}$$

be the singular value decomposition (SVD) of W [14]. We set $N_0 = M$ and define $Y_2 = \hat{V}$, where $\hat{V}$ is the $M \times M$ matrix from the SVD of W.

In this case

$$X = WY = \hat{U}\hat{S}, \tag{70}$$

i.e., the matrix X is formed by eigentrajectories: $X = [s_1\mathbf{u}_1, \ldots, s_M\mathbf{u}_M]$.

Note that Y_1 and Y_2 are connected by an orthogonal transformation in R^M.

3) We set $N_0 = N$ and define $Y_3 = W^*$. Then (65) becomes $X = WW^*$. In this case X is a positive symmetric $M \times M$ matrix. The principal component analysis (PCA) of this matrix (related to the SVD of W (69)) gives

$$X = \sum_{j \in J(r)} \hat{s}_k^2 \hat{\mathbf{u}}_k \hat{\mathbf{u}}_k^*, \tag{71}$$

where r is the rank of the matrix X.

7. Consistent Reconstruction

We now consider how to construct an optimal trajectory by using the data predictive matrix X.

We introduce a target vector (for $t = 0$) $\mathbf{d} \in R^N$. (In general case it is a target subspace with the corresponding metric, see section 2 and 3.)

Note that the vector $\mathbf{d}$ is always outside of the feasible subspace $S = S_x$ due to $t' > t$.

Therefore it is natural to define the orthogonal projection of $\mathbf{d} \in R^N$ onto the feasible subspace S_x:

$$\mathbf{d}_S = P_S \mathbf{d} \tag{72}$$

and consider it as the feasible target for CS. From (72) follows that

$$(\mathbf{x}, \mathbf{d}) = (\mathbf{x}, \mathbf{d}_S), \ \mathbf{x} \in R^N.$$

Indeed, if $\mathbf{x} \in S_x$, then $\mathbf{x} = P_S\mathbf{x}$ and $(\mathbf{x}, \mathbf{d}) = (P_S\mathbf{x}, \mathbf{d}) = (\mathbf{x}, P_S\mathbf{d}) = (\mathbf{x}, \mathbf{d}_S)$. In frame theory the construction of $\mathbf{d}_S$ for a given signal $\mathbf{d}$ is called consistent reconstruction [15], [16].

The vector $\mathbf{d}_S$ minimizes the distance between the vector $\mathbf{d}$ and vectors of the subspace S_x in the metric of the trajectory space R^N. Other metrics (including the metric of the feasible subspace S_x) can also be used for the definition of the real (feasible) target.

Thus, our goal is to construct a feasible trajectory $\mathbf{d}_S$ by using the frame (the data predictive matrix) X. For this we follow constructions described in section 4.

We define the operator

$$G = G_X = XX^*. \tag{73}$$

The operator G maps any trajectory in the trajectories space R^N onto the feasible subspace S_x, because

$$\mathbf{x}' = \sum_{j \in J(N_0)} \mathbf{x}_j \mathbf{x}_j^* \mathbf{x} \tag{74}$$

and the trajectories $\mathbf{x}_j$ span the subspace S_x.

Let

$$X = USV^* \tag{75}$$

be the SVD of the data predictive matrix X.

Then from (73) and (75) we obtain

$$G = US^2U^* = \sum_k s_k^2 \mathbf{u}_k \mathbf{u}_k^*, \tag{76}$$

where the vectors $\mathbf{u}_k \in R^N$ form the orthogonal matrix U.

By using these vectors we can define the orthogonal projector onto the subspace S_x:

$$P_S = \sum_k \mathbf{u}_k \mathbf{u}_k^*. \tag{77}$$

From (76) follows that

$$GG^{-1} = \sum_k \mathbf{u}_k \mathbf{u}_k^* = P_S, \tag{78}$$

where $P_S = \mathrm{Id}_S$ and $\mathbf{x} = GG^{-1}\mathbf{x}$ for all $\mathbf{x} \in S$.

Therefore, for any trajectory $\mathbf{x} \in \mathbf{R}^{\mathbf{N}}$ we have

$$\mathbf{x}_S = P_S\mathbf{x} = \sum_{j \in J(N_0)} \mathbf{x}_j \mathbf{x}_j^* G^{-1}\mathbf{x} = \sum_{j \in J(N_0)} \mathbf{x}_j \tilde{\mathbf{x}}_j^* \mathbf{x}, \tag{79}$$

where

$$\tilde{\mathbf{x}}_j = G^{-1}\mathbf{x}_j.$$

Thus, by using frame X and its dual $\tilde{X}$ we can express the orthogonal projector (77) as $P_S = X\tilde{X}^*$.

Finally, we obtain from (79) for $\mathbf{x} = \mathbf{d}$:

$$\mathbf{d}_S = \sum_{j \in J(N_0)} \mathbf{x}_j \tilde{\mathbf{x}}_j^* \mathbf{d}, \tag{80}$$

because $(\tilde{\mathbf{x}}_j^*, \mathbf{d}) = (\mathbf{x}_j, \mathbf{d})$.

Thus, by estimating the target trajectory $\mathbf{d}$ with respect to the dual frame $\tilde{X}$, CS realizes the trajectory $\mathbf{d}_S$ which is the closest to the target trajectory in the least-square sense.

In frame theory the feasible subspace is considered as a reconstruction space and its dual subspace as a sampling (measurement) space (or vice versa). In [16] a general framework for sampling and reconstruction procedures based on the consistency requirement [15] was considered. It is easy to extend (73)– (80) to this general case.

8. Conclusion

In this chapter we have described a new approach to model predictive control (MPC) based on frame theory and called data predictive control (DPC).

Of course, advantages and drawbacks of this approach can be seen after some period of time. We now can only single out the principal features of this approach:

because of the substitution of a mathematical model by a data matrix there is no need to construct a complicated mathematical model;

the approach handles nonlinear and linear control systems in the same way;

a feasible trajectory is obtained by construction. (However, it may be difficult to construct trajectories that satisfy constraints on the state (output) variables).

References

[1] Mayne, D. Q., and Rawlings, J. B. (2009), *Model Predictive Control: Theory and Design*, WI: Nob Hill Publishing, 576 pp.

[2] Allgower F., Findeisen R., and Ebenbauer C. (2004), Nonlinear model predictive control. *Encyclopedia for Life Support Systems* (EOLSS) article contribution 6.43.16.2, 2004.

[3] Goodwin, G. C., Graebe, S. F., and Salgado, M. E. (2001), *Control System Design*. Englewood Cliffs, NJ: Prentice-Hall.

[4] Propoi, A. I. (1963), Use of linear programming methods for synthesizing sampled-data automatic systems. *Automation and Remote Control*, 24(7), 837-844.

[5] Hughes, R.I.G. (1992), *The Structure and Interpretation of Quantum Mechanics*. Cambridge, MA: Harvard University Press.

[6] Propoi, A. I. (2012), Geometry of model predictive control. *4th IFAC Nonlinear Model Predictive Control Conference*, Noordwijkerhout, The Netherlands, 274-281.

[7] Christensen, O. (2003), *An Introduction to Frames and Riesz Bases*. Boston, MA, U.S.A.: Birkhauser.

[8] Casazza, P.G. (2000), The Art of Frame Theory, *Taiwanese Journal of Math.* 4, No. 2, 1-127.

[9] Goodwin, G. C., Derpich, M.S., and Quevedo, D.E. (2006), Efficient Data Representations for Signal Processing and Control: Making Most of a Little. *Proceedings of the 25th Chinese Control Conference*, Harbin, Heilongjiang.

[10] Beurling, A., and Livingston, A. E. (1962), A theorem on duality mappings in Banach spaces, *Ark. Mat*, 4, 405-411.

[11] Browder, F. E. (1965), Multivalued monotone nonlinear mappings and duality mappings in Banach space, *Trans. Amer. Math. Soc*, 118, 338-351.

[12] Propoi, A. I. (2015), The feasible set in model predictive control. *Chapter in this volume.*

[13] Eldar, Y.C., and Forney, G. D., Jr. (2002), Optimal tight frames and quantum measurement, *IEEE Trans. on Information Theory*, 48, No. 3, 599–610.

[14] Horn, R.A., and Johnson, C.R. (1985), *Matrix Analysis*, Cambridge University Press, Cambridge.

[15] Unser M., and Aldroubi, A. (1994), A general sampling theory for nonideal acquisition devices, *IEEE Trans. Signal Processing*, 42, no. 11, pp. 2915-2925.

[16] Eldar, Y. C. (2003), Sampling with arbitrary sampling and reconstruction spaces and oblique dual frame vectors, *J. Fourier Anal. Appl.*, 9, no. 1, pp. 77-96.

In: Model Predictive Control
Editor: Corrine Wade, pp. 117-143

ISBN: 978-1-63463-859-3
© 2015 Nova Science Publishers, Inc.

Chapter 6

COORDINATED FLEXIBLE DISTRIBUTED MODEL PREDICTIVE CONTROL

Yi Zheng* and Shaoyuan Li
School of Electronic Information and Electrical Engineering,
Shanghai Jiao Tong University, Shanghai, China

Abstract

Consider a complex large-scale control system which is composed of many spatially distributed subsystems. Each subsystem interacts with some other subsystems by their states and/or inputs, e.g. large-scale chemical process, smart grid, distributed generation systems. The control objective is to achieve a specific global performance of the entire system or a common goal of all subsystems.

To control this class of system, the Distributed Model Predictive Control (DMPC), which controls each subsystem by a separate local Model Predictive Control (MPC), has become more and more popular since it not only inherits MPC's ability to explicitly accommodate constraints but also possesses the advantages of the distributed framework of good flexibility and good error tolerance. On the other hand, with the development of communication network technologies in process industries, which allow a distributed controller to access and send information throughout the system, also helps to promote distributed control solutions. However, as point in many articles, the performance of a DMPC is, in most cases, not as good as that of a centralized MPC.

The flexibility (or error tolerance) and global performance is two important characteristics of a DMPC. To improve the optimization performance, the existing methods usually increase the coordination degree (the range of cost that each subsystem-based MPC minimized). With the increasing of the coordination degree, the performance of entire system becomes better and better. However, with the increasing of the coordination degree, the network connectivity become more and more complicity, and consequently the error tolerance and high flexibility become weaker and weaker. It is not expected. Can we find a method which could improve the global performance or coordination degree without any increasing of network connectivity?

In this Chapter, a novel coordination strategy, where each subsystem-based model predictive control (MPC) added a quadratic function of the affection of the current subsystem's input to its down-stream neighbors into its optimization index, is proposed

*E-mail address: yizheng@sjtu.edu.cn

for improving the optimization performance of entire system. This method is able to increase the coordination degree without any increasing of network connections comparing to the methods which do not use this coordination strategy. The consistency constraints, which limit the error between the state predicted at the previous time instant, referred to as the presumed state, and the state predicted at the current time instant within a prescribed bound, are designed and included in the optimization problem of each subsystem-based MPC. These constraints guarantee the recursive feasibility of each subsystem-based MPC. In the meantime, a stabilization constraint and the dual mode predictive control strategy are adopted to result in a stabilizing DMPC.

PACS: 05.45-a, 52.35.Mw, 96.50.Fm

Keywords: Predictive Control, distributed systems, distributed model predictive control, networked control

AMS Subject Classification: 53D, 37C, 65P

1. Introduction

Consider a complex large-scale control system which is composed of many physically or geographically divided subsystems. Each subsystem interacts with some other subsystems by their states and/or inputs, e.g. large-scale chemical process [7], smart grid [8], distributed generation systems. The control objective is to achieve a specific global performance of the entire system or a common goal of all subsystems.

In controlling such a large-scale system, the distributed (or decentralized) framework, where each subsystem is controlled by an independent controller, is usually adopted despite the resulting global performance is in general not as good as a centralized solution [28, 27, 14] for several reasons. The classical centralized control solution is often impractical because of its heavy computational demand and its lack of tolerance to faults in controller. The integrity of the control system cannot be maintained when a control component fails and the whole system is out of control when the centralized controller fails. The distributed framework, in contrast, has the advantages of fault-tolerance, less computation and being flexible to system structure. In the mean time, the development of communication network technologies in process industries, which allows a distributed controller to access and send information throughout the system, also helps to promote distributed control solutions [32].

Among the distributed solutions, the Distributed Model Predictive Control (DMPC), which controls each subsystem by a separate local Model Predictive Control (MPC), has become more and more popular [25, 7] since it not only inherits MPC's ability to explicitly accommodate constraints [24, 19, 17, 30, 35] but also possesses the advantages of the distributed framework mentioned above. However as pointed in [4, 28], the performance of a DMPC is, in most cases, not as good as that of a centralized MPC.

Many algorithms have appeared in the literature for different type of systems and for different problems in the design of DMPC, e.g. design of DMPC for nonlinear systems [2, 12], design DMPC for networked systems with time delay [18, 36], development of decentralized optimization algorithm for DMPC [6], and design cooperative strategy for improving performance of DMPC [33, 9], as well as the design of control structure of DMPC [1]. Among them, several coordination strategies have appeared to improve the global performance of the DMPC, and can be classified according to the information exchange protocol needed (i.e non-iterative or iterative algorithms), and to the type of cost function which is optimized [25]. The DMPCs, which accommodate same kind of cost function for each subsystem-based MPC, can be solved either by iterative algorithm or non-iterative algorithm. Thus the coordination strategies are introduced by the classification of the type of cost function which is optimized in each subsystem-based MPC. The simplest and most adopted strategy is that each local controller minimizes its own subsystem's cost and uses the state prediction of the previous time instant to approximate the state sequence at the current time instant in computing the optimal solution [4, 15, 10]. In this kind of method, each subsystem based controller should communicate with its neighbors. Another commonly used coordination strategy is that each subsystem-based MPC optimizes the cost of overall system to improve the global performance [28, 26, 33, 9, 31]. In computing the optimal solution, it also uses the state prediction of the previous time instant to approximate the state sequence at the current time instant. This strategy could achieve a good global performance in some cases, but it reduces the flexibility and increases the communication load since each subsystem-based MPC should communicate with all other subsystems. In an effort to achieve a trade off between the global performance of the entire system and the computational burden, recently, an intuitively appealing strategy is proposed in [34, 32, 36, 16], where each subsystem-based MPC only considers the cost of its own subsystem and those of the subsystems it directly impacts on. Such a design can be referred to as *Impacted-region Cost Optimization based DMPC* (ICO-DMPC). In particular, Ref. [34, 36] applies this design idea to a metallurgy system and [32] explains why this coordination strategy could improve the global performance. Numerical and practical experiments show that this coordination strategy could obtain a performance close to that of a classical centralized MPC. In this kind of method, each subsystem based controller should communicate with its neighbors and its neighbours' neighbours.

It can be seen from above that, in existing methods, with the increasing of the coordination degree, the performance of entire system becomes better and better, the network connectivity becomes more and more complicity, and consequently the error tolerance and high flexibility become weaker and weaker. To find a method which could improve the global performance or coordination degree without any increasing of network connectivity is still remain to be solved.

In this Chapter, an Coordinated Flexible Distributed Model Predictive Control (CF-DMPC) is proposed and presented for distributed systems, where each subsystem-based MPC adds a quadratic function of the impact of current subsystem's input to its down stream neighbours into its optimization index to increase the coordination degree. It do not need any additional network connectivity comparing to the approach without this coordination strategy. The consistency constraints, which limit the error between the state predicted at the previous time instant, referred to as the presumed state, and the state predicted at the current time instant within a prescribed bound, are designed and included in the optimization problem of each subsystem-based MPC. These constraints can bound the error between the presumed state and the predictive state of upstream neighbors and the error between the predictive state of downstream neighbor calculated by current subsystem-based MPC and that of the downstream neighbors themselves. And these constraints guarantee the feasibility of each subsystem-based MPC. They also guarantee that the remaining part of the

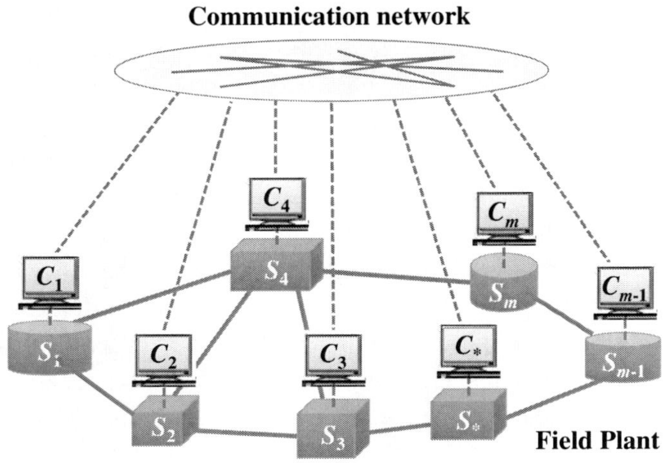

Figure 1. An illustration of the structure of a distributed system and its distributed control framework.

solution at the previous time instant is a feasible solution. In the mean time, a stabilization constraint and the dual mode predictive control [13, 23] strategy are adopted to result in a stabilizing ICO-DMPC.

The remainder of this Chapter is organized as follows. Section 2 describes the problem to be solved in this paper. Section 3 presents the design of the stabilizing CF-DMPC. The feasibility of the proposed CF-DMPC and the stability of the resulting closed-loop system are analyzed in Section 4. Section 5 presents the simulation results to demonstrate the effectiveness of the proposed DMPC algorithm. Finally, a brief conclusion to the paper is drawn in Section 6.

2. Problem Description

2.1. Distributed Systems

A distributed system, as illustrated in Fig. 1, is composed of many interacting subsystems, each of which is controlled by a subsystem-based controller, which in turn is able to exchange information with other subsystem-based controllers.

Suppose that the distributed system $\mathcal{S}$ is composed of m discrete-time linear subsystems $\mathcal{S}_i$, $i \in \mathcal{P} = \{1, 2, \cdots, m\}$ and m controllers $\mathcal{C}_i$, $i \in \mathcal{P} = \{1, 2, \cdots, m\}$. Let the subsystems interact with each other through their states. If subsystem $\mathcal{S}_i$ is affected by $\mathcal{S}_j$, for any $i \in \mathcal{P}$ and $j \in \mathcal{P}$, subsystem $\mathcal{S}_i$ is said to be a downstream system of subsystem $\mathcal{S}_j$, and subsystem $\mathcal{S}_j$ is an upstream system of $\mathcal{S}_i$. Let $\mathcal{P}_{+i}$ denote the set of the subscripts of the upstream systems of $\mathcal{S}_i$, $\mathcal{P}_{-i}$ is the set of the subscripts of the downstream systems of $\mathcal{S}_i$. Then, subsystem $\mathcal{S}_i$ can be expressed as

$$\begin{cases} \mathbf{x}_{i,k+1} = \mathbf{A}_{ii}\mathbf{x}_{i,k} + \mathbf{B}_{ii}\mathbf{u}_{i,k} + \sum_{j \in \mathcal{P}_{+i}} \mathbf{A}_{ij}\mathbf{x}_{j,k}, \\ \mathbf{y}_{i,k} = \mathbf{C}_{ii}\mathbf{x}_{i,k}, \end{cases} \tag{1}$$

where $\mathbf{x}_i \in \mathbf{R}^{n_{xi}}$, $\mathbf{u}_i \in \mathcal{U}_i \subset \mathbf{R}^{n_{ui}}$ and $\mathbf{y}_i \in \mathbf{R}^{n_{yi}}$ are respectively the local state, input and output vectors, and $\mathcal{U}_i$ is the feasible set of the input $\mathbf{u}_i$, which is used to bound the input according to the physical constraints on the actuators, the control requirements or the characteristics of the plant. A non-zero matrix $\mathbf{A}_{ij}$, that is, $j \in \mathcal{V}_i^-$, indicates that $\mathcal{S}_i$ is affected by $\mathcal{S}_j$. In the concatenated vector form, the system dynamics can be written as

$$\begin{cases} \mathbf{x}_{k+1} &= \mathbf{A}\mathbf{x}_k + \mathbf{B}\mathbf{u}_k, \\ \mathbf{y}_k &= \mathbf{C}\mathbf{x}_k, \end{cases} \tag{2}$$

where $\mathbf{x} = [\mathbf{x}_1^{\mathrm{T}}\ \mathbf{x}_2^{\mathrm{T}}\ \cdots\ \mathbf{x}_m^{\mathrm{T}}]^{\mathrm{T}} \in \mathbf{R}^{n_x}$, $\mathbf{u} = [\mathbf{u}_1^{\mathrm{T}}\ \mathbf{u}_2^{\mathrm{T}}\ \cdots\ \mathbf{u}_m^{\mathrm{T}}]^{\mathrm{T}} \in \mathbf{R}^{n_u}$ and $\mathbf{y} = [\mathbf{y}_1^{\mathrm{T}}\ \mathbf{y}_2^{\mathrm{T}}\ \ldots\ \mathbf{u}_m^{\mathrm{T}}]^{\mathrm{T}} \in \mathbf{R}^{n_y}$ are respectively the concatenated state, control input and output vectors of the overall system $\mathcal{S}$, and A, B and C are constant matrices of appropriate dimensions. Also, $\mathbf{u} \in \mathcal{U} = \mathcal{U}_1 \times \mathcal{U}_2 \times \cdots \times \mathcal{U}_m$.

The control objective is to stabilize the overall system $\mathcal{S}$ in an DMPC framework. Meanwhile, the achieved performance index of the overall system should be as close as possible to the performance index achievable under a centralized MPC and the network connectivity should not increase.

2.2. Existing Methods and Motivations

The works on DMPCcan be roughly divided in optimization-based frameworks which focus on the solution of the optimization problems [21]. and control- based frameworks which are concerned mostly with stability and control performance [5].

The control-based frameworks can be classified according to the information exchange protocol needed (i.e non-iterative or iterative algorithms), and to the type of cost function which is optimized. The non-iterative algorithms which only communicates once a control period have faster computational speed comparing to that of iterative algorithm and the iterative algorithms could obtains more excellent optimization performance than non-iterative algorithms. The DMPC which accommodate same kind of cost function for subsystem-based MPC can be solved either by iterative algorithm or non-iterative algorithm. Different methods have different characteristics and are suitable for different control purposes. We briefly review these methods as motivations for the problem we have just formulated and its solution to be presented later in the paper.

1) Distributed algorithms where each subsystem-based controller minimizes the cost function of its own subsystem were proposed in [4, 15],

$$J_{i,k} = \|\mathbf{x}_{i,k+N}\|^2_{\mathbf{P}_i} + \sum_{l=0}^{N-1} \left(\|\mathbf{x}_{i,k+l}\|^2_{\mathbf{Q}_i} + \|\mathbf{u}_{i,k+l}\|^2_{\mathbf{R}_i} \right) \tag{3}$$

When computing the optimal solution, each local controller exchanges state estimation with the neighboring subsystems to improve the performance of the local subsystem. This method is simple and very convenient for implementation. An extension of this stabilizing DMPC with input constraint for nonlinear continuous systems is given in [10]. In addition [11] gives a design for linear systems, which could handle both input and states constraints by using a fixed reference state trajectory with moving widow to replace the state estimation in each update.

2) To improve the global performance, distributed algorithms, where each local controller minimizes a global cost function

$$\widetilde{J}_{i,k} = \sum_{j \in \mathcal{P}} J_{j,k}, \tag{4}$$

were proposed in [28, 33, 26, 9, 31]. In this method, each subsystem-based MPC exchanges information (the estimation of the future state sequences of the subsystems) with all other subsystems. And some iterative stabilizing designs are proposed which take the advantages of the model of whole system is used in each subsystem-based MPC. This strategy may result in a better performance but consumes much more communication resources, in comparison with the method in described in 1).

3) To balance the performance, communication cost and the complexity of the DMPC algorithm, the strategy that each subsystem-based controller only minimizes its own cost function and those of the subsystems that its own subsystem directly impacts on was recently proposed in [34, 32, 36, 16], that is

$$\bar{J}_{i,k} = \sum_{j \in \mathcal{P}_i} J_{j,k}, \tag{5}$$

where $\mathcal{P}_i = \{j : j \in \mathcal{P}_{-i} \text{ or } j = i\}$ is the set of subscripts of the downstream subsystems of subsystem $\mathcal{S}_i$, that is the region impacted on by subsystem $\mathcal{S}_i$. The resulting control algorithm is termed as an Impacted-region Cost Optimization based DMPC, or ICO-DMPC. It could achieve a better performance than the first method, and its communication burden is much less than the second method [1, 3].

Clearly, this coordination strategy as proposed in [34, 32] and described in 3) is a preferable method to trade-off the communication burden and the global performance. However, the DMPC under this coordination strategy the network conductivities is still increase, although the global performance is dramatically increased with less increasing of network connectivity. A method which could increase the coordination degree of DMPC without increasing of network connectivity remains to be developed. The objective of this Chapter is to develop and introduce such a DMPC design.

3. Coordinated Flexible Distributed Model Predictive Control Formulation

In this Section, m separate optimal control problems, one for each subsystem, and the Coordinated Flexible Distributed Model Predictive Control (CF-DMPC) which coordinates the subsystem-based MPCs by adding a quadratic function of the impact of the current subsystem's input to its down stream neighbours into its optimization index is defined. In every distributed optimal control problem, the same constant prediction horizon N, $N > 1$, is used. And every distributed MPC law is updated globally synchronously. At each update, every local MPC optimizes only for its own open-loop control sequence, given the current states and the estimated inputs of the whole system.

To proceed, we need the following assumption,

Table 1. Notation

Notation	Explanation
$\mathcal{P}$	the set of the subscripts of all subsystems
$\mathcal{P}_i$	the set of the subscripts of all subsystems, excluding $\mathcal{S}$ itself
$+i$	the subscript denote all upstream neighbors of $\mathcal{S}_i$
$-i$	the subscript denote all downstream neighbors of $\mathcal{S}_i$
$\mathbf{u}_{i,k+l-1\|k}$	the optimal control sequence of $\mathcal{S}_i$ calculated by $\mathcal{C}_i$ at time k
$\hat{\mathbf{x}}_{i,k+l\|k}$	the presumed state sequence of $\mathcal{S}_i$ define by $\mathcal{C}_i$ at time k
$\hat{\mathbf{u}}_{i,k+l\|k}$	the presumed input sequence of $\mathcal{S}_i$ define by $\mathcal{C}_i$ at time k
$\mathbf{u}^{\mathrm{f}}_{i,k+l-1\|k}$	the feasible control at time k+l-1 of $\mathcal{S}_i$ defined by $\mathcal{C}_i$ at time k
$\mathbf{x}^{\mathrm{f}}_{i,k+l\|k}$	the predictive feasible state sequence of $\mathcal{S}_i$ defined by $\mathcal{C}_i$ at time k
$\mathbf{u}^{\mathrm{p}}_{i,k+l-1\|k}$	the predicted control at time k+l-1 of $\mathcal{S}_i$ defined by $\mathcal{C}_i$ at time k
$\mathbf{x}^{\mathrm{p}}_{i,k+l\|k}$	the predicted state sequence of $\mathcal{S}_i$ defined by $\mathcal{C}_i$ at time k
$\|\|\cdot\|\|_{\mathbf{P}}$	refer to the P norm, P is any positive matrix, and $\|\|\mathbf{z}\|\|_{\mathbf{P}} = \sqrt{\mathbf{x}_{\mathbf{k}}^{\mathrm{T}}\mathbf{P}\mathbf{x}_{\mathbf{k}}}$.

Assumption 1. *For every subsystem $\mathcal{S}_i$, $i \in \mathcal{P}$, there exists a decoupled static feedback $\mathbf{u}_i = \mathbf{K}_i\mathbf{x}_i$ such that $\mathbf{A}_{\mathrm{d}i} = \mathbf{A}_{ii}+\mathbf{B}_{ii}\mathbf{K}_i$ is Shur stable, and the closed-loop system $\mathbf{x}_{k+1} = \mathbf{A}_{\mathrm{c}}\mathbf{x}_k$ is asymptotically stable, where $\mathbf{A}_{\mathrm{c}} = \mathbf{A} + \mathbf{B}\mathbf{K}$ and $\mathbf{K} = \mathrm{diag}\{\mathbf{K}_1, \mathbf{K}_2, \cdots, \mathbf{K}_m\}$.*

This assumption is usually used in the design of stabilizing DMPC, see [11, 10]. It presumes that each subsystem is able to be stabilized by a decentralized control $\mathbf{K}_i\mathbf{x}_i$, $i \in \mathcal{V}$. and the decentralized control gain $\mathbf{K}$ based on LMI have been proposed in [29] for continuous time systems. These techniques can be easily adapted to the discrete-time case here considered.

We also define the necessary notation in Table 1.

Consider that the control law of current subsystem $\mathcal{S}_i$ effects the performance of its downstream neighboring subsystems $\mathcal{S}_j$, $j \in \mathcal{P}_{-i}$, in the CF-DMPC, the performance of $\mathcal{S}_j$ is added into the performance index of the MPC which control $\mathcal{S}_i$ based on a approximation of the updated state sequence of $\mathcal{S}_j$. The approximated state sequence equals the assumed state sequence of $\mathcal{S}_j$ pluses the impact caused by the change of control law of $\mathcal{S}_i$ to the state sequence of $\mathcal{S}_j$. In this way, the coordination degree is expanded without any increasing of the required network connectivity in solving each subsystem-based MPC.

Define that $\mathbf{f}_{i,k+l|k}$ be the matching from $\mathbf{u}_{i,k:k+l-1|k}$ to $\mathbf{x}_{i,k+l|k}$, and it can be deduced from equation (1) as

$$\begin{aligned}\mathbf{f}_{i,k+l|k} &= \mathbf{x}_{i,k+l|k} \\ &= \mathbf{A}_{ii}^{l}\mathbf{x}_{i,k} + \sum_{h=1}^{l} \mathbf{A}_{ii}^{l-h}\mathbf{B}_{ii}\mathbf{u}_{i,k+h-1|k} \\ &\quad + \sum_{j\in\mathcal{P}_{+i}} \sum_{h=1}^{l} \mathbf{A}_{ii}^{l-h}\mathbf{A}_{ij}\mathbf{x}_{j,k+h-1|k} \end{aligned} \tag{6}$$

Then, it have

$$\frac{\partial \mathbf{f}_{i,k+l|k}}{\partial \mathbf{x}_{j,k+h-1|k}} = \mathbf{A}_{ii}^{l-h}\mathbf{A}_{ij} \tag{7}$$

$$\frac{\partial \mathbf{x}_{i,k+l|k}}{\partial \mathbf{u}_{i,k+h-1|k}} = \mathbf{A}_{ii}^{l-h}\mathbf{B}_{ii} \tag{8}$$

The $\mathbf{f}_{i,k+l|k}$ derivation of the $\mathbf{u}_{j,k+h-1|k}$ becomes

$$\begin{aligned}\frac{\partial \mathbf{f}_{i,k+l|k}}{\partial \mathbf{u}_{j,k+h-1|k}} &= \sum_{p=h+1}^{l} \frac{\partial \mathbf{f}_{i,k+l|k}}{\partial \mathbf{x}_{j,k+p-1|k}} \frac{\partial \mathbf{x}_{j,k+p-1|k}}{\partial \mathbf{u}_{j,k+h-1|k}} \\ &= \sum_{p=h+1}^{l} \mathbf{A}_{ii}^{l-p}\mathbf{A}_{ij}\mathbf{A}_{jj}^{p-h}\mathbf{B}_{jj}. \end{aligned} \tag{9}$$

Since the state and input sequences of downstream and upstream neighbors of $\mathcal{S}_i$ is unknown to the controller of $\mathcal{S}_i$, assume that the state and input sequences $\hat{\mathbf{x}}_{i,k+l|k}$ and $\hat{u}_{i,k+l|k}$ be the presumed states and presumed input which are calculated in the previous calculation, respectively. Add the estimation of the performance of the $\mathcal{S}_j$, $j \in \mathcal{P}_{-i}$ to the cost function of the MPC for $\mathcal{S}_i$, then the optimization index of $\mathcal{S}_i$ becomes

$$\begin{aligned}\bar{J}_i(k) &= \sum_{l=1}^{N} \left(\left\| \mathbf{x}_{i,k+l|k}^{\mathrm{P}} \right\|_{\mathbf{Q}_i}^2 + \left\| \mathbf{u}_{i,k+l-1|k} \right\|_{\mathbf{R}_i}^2 \right) \\ &\quad + \sum_{j\in\mathcal{P}_{-i}} \sum_{l=1}^{N} \left\| \left(\hat{\mathbf{x}}_{j,k+l|k} + \omega_i \mathbf{S}_{ji,k+l|k} \right) \right\|_{\mathbf{Q}_j}^2 + \sum_{j\in\mathcal{P}_{-i}} \sum_{l=1}^{N} \left\| \hat{\mathbf{u}}_{i,k+l-1|k} \right\|_{\mathbf{R}_j}^2 \end{aligned} \tag{10}$$

where ω_i is the weighting coefficients for improving the convergence when using iterative algorithm, and

$$\begin{aligned}&\mathbf{S}_{ji,k+l|k} = \sum_{h=1}^{l} \sum_{p=h+1}^{l} \mathbf{A}_{jj}^{l-p}\mathbf{A}_{ji}\mathbf{A}_{ii}^{p-h}\mathbf{B}_{ii}\left(\mathbf{u}_{i,k+h-1|k} - \hat{\mathbf{u}}_{i,k+h-1|k}\right) \\ &h = 1, 2, \ldots, l. \end{aligned} \tag{11}$$

where $\mathbf{Q}_i = \mathbf{Q}_i^{\mathrm{T}} > 0$, $\mathbf{R}_i = \mathbf{R}_i^{\mathrm{T}} > 0$ and $\mathbf{P}_j = \mathbf{P}_j^{\mathrm{T}} > 0$. The matrix $\mathbf{P}_i$ is chosen to satisfy the Lyapunov equation

$$\mathbf{A}_{\mathrm{d}i}^{\mathrm{T}}\mathbf{P}_i\mathbf{A}_{\mathrm{d}i} - \mathbf{P}_i = -\widehat{\mathbf{Q}}_i,$$

where $\widehat{\mathbf{Q}}_i = \mathbf{Q}_i + \mathbf{K}_i^{\mathrm{T}}\mathbf{R}_i\mathbf{K}_i$. Denote

$$\begin{aligned}
\mathbf{P} &= \operatorname{diag}\{\mathbf{P}_1, \mathbf{P}_2, \cdots, \mathbf{P}_m\},\\
\mathbf{Q} &= \operatorname{diag}\{\mathbf{Q}_1, \mathbf{Q}_2, \ldots, \mathbf{Q}_m\},\\
\mathbf{R} &= \operatorname{diag}\{\mathbf{R}_1, \mathbf{R}_2, \cdots, \mathbf{R}_m\},\\
\mathbf{A}_{\mathrm{d}} &= \operatorname{diag}\{\mathbf{A}_{\mathrm{d}1}, \mathbf{A}_{\mathrm{d}2}, \cdots, \mathbf{A}_{\mathrm{d}m}\}.
\end{aligned}$$

Then, it follows that

$$\mathbf{A}_{\mathrm{d}}^{\mathrm{T}}\mathbf{P}\mathbf{A}_{\mathrm{d}} - \mathbf{P} = -\widehat{\mathbf{Q}},$$

where $\widehat{\mathbf{Q}} = \mathbf{Q} + \mathbf{K}^{\mathrm{T}}\mathbf{R}\mathbf{K} > 0$.

Since every subsystem-based controller is updated synchronously, the state and control sequences of other subsystems are unknown to subsystem $\mathcal{S}_i$. Thus, at the time instant k, the presumed state sequence $\hat{\mathbf{x}}_{j,k:k+N|k}$ of $\mathcal{S}_j$ are used in the predictive model of the MPC in $\mathcal{S}_i$, which is given as

$$\begin{aligned}
\mathbf{x}^{\mathrm{p}}_{i,k+l|k} &= \mathbf{A}_{ii}^{l}\mathbf{x}^{\mathrm{p}}_{i,k} + \sum_{h=1}^{l}\mathbf{A}_{ii}^{l-h}\mathbf{B}_{ii}\mathbf{u}_{i,k+h-1|k}\\
&\quad + \sum_{j\in\mathcal{P}_i^u}\sum_{h=1}^{l}\mathbf{A}_{ii}^{l-h}\mathbf{A}_{ij}\hat{\mathbf{x}}_{j,k+h-1|k}
\end{aligned} \tag{12}$$

Given $\mathbf{x}^{\mathrm{p}}_{i,k|k} = \mathbf{x}_i(k|k)$, the presumed control sequence for subsystem $\mathcal{S}_i$ is given by

$$\hat{\mathbf{u}}_{i,k+l-1|k} = \begin{cases} \mathbf{u}^{\mathrm{p}}_{i,k+l-1|k-1}, l = 1, 2, \cdots, N-1, \\ \mathbf{K}_i\mathbf{x}^{\mathrm{p}}_{i,k+N-1|k-1}, l = N \end{cases} \tag{13}$$

Set each presumed state sequence $\hat{\mathbf{x}}_i$ to be the remainder of the sequence predicted at the previous time instant $k-1$, concatenated with the closed-loop response under the state feedback control $\hat{\mathbf{u}}_{i,k+l-1|k} = \mathbf{K}_j\mathbf{x}^{\mathrm{p}}_{i,k+N-1|k-1}$, that is that is

$$\begin{cases} \hat{\mathbf{x}}_{i,k+l-1|k} = \mathbf{x}^{\mathrm{p}}_{k+l-1|k-1}, l = 1, 2, \cdots, N, \\ \hat{\mathbf{x}}_{i,k+N+1-1|k} = \mathbf{A}_{\mathrm{d}i}\mathbf{x}^{\mathrm{p}}_{i,k+N-1|k-1} + \sum_{j\in\mathcal{P}_{+i}}\mathbf{A}_{ij}\mathbf{x}^{\mathrm{p}}_{j,k+N-1|k-1} \end{cases} \tag{14}$$

In MPC systems, it is an important proposition to focus on the systems sequence feasibility and the stability. The same things happen in the distributed MPC systems. To enlarge the feasible region, a terminal state constraint is included in each subsystem-based MPC, which guarantees that the terminal controllers are stabilizing inside a terminal set. To define this terminal state set, we need to make an assumption and establish a technical lemma.

Assumption 2. *The block-diagonal matrix* $\mathbf{A}_{\mathrm{d}} = \operatorname{diag}\{\mathbf{A}_{\mathrm{d}1}, \mathbf{A}_{\mathrm{d}2}, \ldots, \mathbf{A}_{\mathrm{d}m}\}$ *and the off-diagonal matrix* $\mathbf{A}_{\mathrm{o}} = \mathbf{A}_{\mathrm{c}} - \mathbf{A}_{\mathrm{d}}$ *satisfy the following inequality*

$$\mathbf{A}_{\mathrm{o}}^{\mathrm{T}}\mathbf{P}\mathbf{A}_{\mathrm{o}} + \mathbf{A}_{\mathrm{o}}^{\mathrm{T}}\mathbf{P}\mathbf{A}_{\mathrm{d}} + \mathbf{A}_{\mathrm{d}}^{\mathrm{T}}\mathbf{P}\mathbf{A}_{\mathrm{o}} < \widehat{\mathbf{Q}}/2.$$

It along with Assumption 1., help with the design of the terminal set. This assumption quantifies the coupling between subsystems which is sufficiently weak to be controlled by the algorithm designed here.

Lemma 3.1. *Under Assumptions 1. and 2., for any positive scalar c the set*

$$\Omega(c) = \{\mathbf{x} \in \mathbf{R}^{n_x} : \|\mathbf{x}\|_{\mathbf{P}} \leq c\}$$

is a positive invariant region of attraction for the closed-loop system $\mathbf{x}_{k+1} = \mathbf{A}_{\mathrm{c}}\mathbf{x}_k$. Additionally, there exists a small enough positive scalar ε such that $\mathbf{Kx}$ is in the feasible input set $\mathcal{U} \subset \mathbf{R}^{n_u}$ for all $\mathbf{x} \in \Omega(\varepsilon)$.

Proof. Consider the function $V(k) = \|\mathbf{x}_k\|^2_{\mathbf{P}}$. The time difference of $V(k)$ along the trajectories of the closed-loop system $\mathbf{x}_{k+1} = \mathbf{A}_{\mathrm{c}}\mathbf{x}_k$ can be evaluated as

$$\begin{aligned}
\Delta V_k &= \mathbf{x}_k^{\mathrm{T}}\mathbf{A}_{\mathrm{c}}^{\mathrm{T}}\mathbf{P}\mathbf{A}_{\mathrm{c}}\mathbf{x}_k - \mathbf{x}_k^{\mathrm{T}}\mathbf{P}\mathbf{x}_k \\
&= \mathbf{x}_k^{\mathrm{T}}(\mathbf{A}_{\mathrm{d}}^{\mathrm{T}}\mathbf{P}\mathbf{A}_{\mathrm{d}} - \mathbf{P} + \mathbf{A}_{\mathrm{o}}^{\mathrm{T}}\mathbf{P}\mathbf{A}_{\mathrm{o}} + \mathbf{A}_{\mathrm{o}}^{\mathrm{T}}\mathbf{P}\mathbf{A}_{\mathrm{d}} + \mathbf{A}_{\mathrm{d}}^{\mathrm{T}}\mathbf{P}\mathbf{A}_{o})\mathbf{x}_k \\
&\leq -\mathbf{x}_k^{\mathrm{T}}\hat{\mathbf{Q}}\mathbf{x}_k + \frac{1}{2}\mathbf{x}_k^{\mathrm{T}}\hat{\mathbf{Q}}\mathbf{x}_k \\
&\leq 0,
\end{aligned}$$

which holds for all $\mathbf{x}(k) \in \Omega(c) \setminus \{0\}$. This implies that all trajectories of the closed-loop system that starts inside $\Omega(c)$ will remain inside and converge to the origin.

The existence of an $\varepsilon > 0$ such that $K\mathbf{x} \in \mathcal{U}$ for all $\mathbf{x} \in \Omega(\varepsilon)$ follows from the fact that P is positive definite, which implies that the set $\Omega(\varepsilon)$ shrinks to the origin as ε decreases to zero. This completes the proof. □

In the optimization problem of each subsystem-based MPC, the terminal state constraint set for each $\mathcal{S}_i$ can then be set to be

$$\Omega_i(\varepsilon) = \left\{\mathbf{x}_i \in \mathbf{R}^{n_{xi}} : \|\mathbf{x}_i\|_{\mathbf{P}_i} \leq \varepsilon/\sqrt{m}\right\}.$$

Clearly, if $\mathbf{x} \in \Omega_1(\varepsilon) \times \cdots \times \Omega_m(\varepsilon)$, then the decoupled controllers will stabilize the system at the origin, since

$$\|\mathbf{x}_i\|^2_{\mathbf{P}_i} \leq \frac{\varepsilon^2}{m}, \ \forall i \in \mathcal{P},$$

implies that

$$\sum_{i \in \mathcal{P}} \|\mathbf{x}_i\|^2_{\mathbf{P}_i} \leq \varepsilon^2,$$

which in turn implies that $x \in \Omega(\varepsilon)$. Suppose that at some time k_0, $\mathbf{x}_{i,k_0} \in \Omega_i(\varepsilon)$ for every subsystem. Then, by Lemma 3.1., stabilization can be achieved if every $\mathcal{C}_i$, $i \in \mathcal{P}$, employs its decoupled static feedback controller $K_i\mathbf{x}_{i,k}$ after time instant k_0.

Thus, the objective of each subsystem-based MPC law is to drive the state of each subsystem $\mathcal{S}_i$ to the set $\Omega_i(\varepsilon)$. Once all subsystems have reached these sets, they switch to their decoupled controllers for stabilization. Such switching from an MPC law to a terminal controller once the state reaches a suitable neighborhood of the origin is referred to as the dual-mode MPC [23, 13]. For this reason, the DMPC algorithm we propose in this paper is a dual-mode DMPC algorithm.

In what follows, we formulate the optimization problem for each subsystem-based MPC.

Problem 1. *Consider subsystem* $\mathcal{S}_i$. *Let* $\varepsilon > 0$ *be as specified in Lemma 3.1.. Let the update time be* $k \geq 1$. *Given* $\mathbf{x}_{i,k}$, $bf\hat{x}_{j,k+l|k}$, $l = 1, 2, \ldots, N$, $\forall j \in \mathcal{P}_{+i}$, *find the control sequence* $\mathbf{u}_{i,k+l|k} : \{0, 1, \cdots, N-1\} \rightarrow \mathcal{U}_i$ *that minimizes*

$$\begin{aligned} \bar{J}_i(k) &= \sum_{l=1}^{N} \left(\left\| \mathbf{x}^{\mathrm{p}}_{i,k+l|k} \right\|^2_{\mathbf{Q}_i} + \left\| \mathbf{u}_{i,k+l-1|k} \right\|^2_{\mathbf{R}_i} \right) \\ &\quad + \sum_{j \in \mathcal{P}_{-i}} \sum_{l=1}^{N} \left\| \hat{\mathbf{x}}_{j,k+l|k} + \omega_i \mathbf{S}_{ji,k+l|k} \right\|^2_{\mathbf{Q}_j} + \sum_{j \in \mathcal{P}_{-i}} \sum_{l=1}^{N} \left\| \hat{\mathbf{u}}_{i,k+l-1|k} \right\|^2_{\mathbf{R}_j} \end{aligned} \tag{15}$$

Subject to the constraints:

$$(12), \tag{16}$$

$$\sum\nolimits_{s=1}^{l} \alpha_{l-s} \| \mathbf{x}^{\mathrm{p}}_{i,k+s|k} - \hat{\mathbf{x}}_{i,k+s|k} \|_2 \leq \frac{\xi \kappa \varepsilon}{2\sqrt{m m_1}}, l = 1, 2, \ldots, N-1 \tag{17}$$

$$\left\| \mathbf{x}^{\mathrm{p}}_{i,k+N|k} - \hat{\mathbf{x}}_{i,k+N|k} \right\|_{\mathbf{P}_i} \leq \frac{\kappa \varepsilon}{2\sqrt{m}}, \tag{18}$$

$$\left\| \mathbf{x}^{\mathrm{f}}_{i,k+l|k} \right\|_{\mathbf{P}_i} \leq \left\| \mathbf{x}^{\mathrm{p}}_{i,k+l|k} \right\|_{\mathbf{P}_i} + \frac{\varepsilon}{\mu N \sqrt{m}}, l = 1, 2, \cdots, N, \tag{19}$$

$$\mathbf{u}_{i,k+l|k} \in \mathcal{U}_i, l = 0, 1, \ldots, N-1, \tag{20}$$

$$\mathbf{x}^{\mathrm{p}}_{i,k+N|k} \in \Omega_i(\varepsilon/2) \tag{21}$$

In the constraints above,

$$m_1 = \max_{i \in \mathcal{P}} \{\text{number of elements in } \mathcal{P}_{+i}\}, \tag{22}$$

$$\alpha_l = \max_{i \in \mathcal{P}} \max_{j \in \mathcal{P}_i} \left\{ \lambda^{\frac{1}{2}}_{\max} \left(\left(\mathbf{A}^l_{ii} \mathbf{A}_{ij} \right)^T \mathbf{P}_j \mathbf{A}^l_{ii} \mathbf{A}_{ij} \right) \right\}, l = 0, 1, \cdots, N-1, \tag{23}$$

The constant $0 < \kappa < 1$, $0 < \xi \leq 1$ are design parameters whose values will be chosen in the sequel. Both (17) and (18) are referred to as the consistency constraints, which require that each predictive sequence and control variables remain close to their presumed values. These constraints are keys to proving that $\mathbf{x}^{\mathrm{f}}_i$ is a feasible state sequence at each update.

Equation (19) will be utilized to prove that the CF-DMPC algorithm is stabilizing, where $\mu > 0$ is a design parameter whose value will be specified later. $\mathbf{x}^{\mathrm{f}}_{i,k+l|k}$ is a feasible state sequence, and equals to the solution of (12) under the initial state of $\mathbf{x}_{i,k}$, the feasible control sequence $\mathbf{u}^{\mathrm{f}}_{i,k+l-1|k}$ is defined by

$$\mathbf{u}^{\mathrm{f}}_{i,k+l-1|k} = \begin{cases} \mathbf{u}^{\mathrm{p}}_{i,k+l-1|k-1}, & l = 1, 2, \cdots, N-1, \\ \mathbf{K}_i \mathbf{x}^{\mathrm{f}}_{i,k+N-1|k}, & l = N. \end{cases} \tag{24}$$

It should be noticed that the terminal constraint in each optimal control problem is $\Omega_i(\varepsilon/2)$, although Lemma 3.1. ensures that the larger $\Omega_i(\varepsilon)$ suffices for the feasibility of the terminal controllers. In the analysis presented in the next section, it will be shown that tightening the terminal set in this way is required to guarantee the feasibility properties.

Before stating the CF-DMPC algorithm, an assumption is made to facilitate the initialization phase.

Assumption 3. *At initial time k_0, there exists a feasible control $\mathbf{u}_{i,k_0+l} \in \mathcal{U}_i$, $l \in \{1,\ldots,N\}$, for each $\mathcal{S}_i$, such that the solution to the full system $\mathbf{x}_{l+1+k_0} = \mathbf{A}\mathbf{x}_{l+k_0} + \mathbf{B}\mathbf{u}_{l+k_0}$, denoted $\hat{\mathbf{x}}_{\cdot|k_0,i}$, satisfies $\hat{\mathbf{x}}_{N+k_0|k_0,i} \in \Omega(\alpha\varepsilon)$ and results in a bounded cost $\bar{J}_{i,k_0}$.*

Remark 3.1. *Assumption 3. bypasses the difficult task of actually constructing an initially feasible solution in a distributed way. In fact, finding an initially feasible solution for many optimization problems is often a primary obstacle, whether or not such problems are used in a control setting. As such, many centralized implementations of MPC also assume that an initially feasible solution is available [23, 20]. Recent methods for characterizing the set of initially feasible solutions are presented in [22]. One possible way to obtain an initially feasible solution can be to solve the corresponding centralized MPC solution at the initial time instant.*

The dual-mode FC-MPC law for any $\mathcal{S}_i$, which communicates once every update, is as follows.

Algorithm 1. (CF-DMPC Algorithm)

Step 1: Initialization at time k_0.

- *Initialize $\mathbf{x}_{k_0}$, $u_{i,k_0+l-1|k_0}$, $l = 1, 2, \ldots, N$. to satisfy assumption 3.*
- *At time k_0, if $\mathbf{x}_{k_0} \in \Omega(\varepsilon)$, then apply the terminal controller $\mathbf{u}_{i,k} = \mathbf{K}_i\mathbf{x}_{i,k}$, for all $k \geq k_0$; Else*
- *Compute $\hat{\mathbf{x}}_{i,k_0+l+1|k_0+1}$ according to (12) and transmit $\hat{\mathbf{x}}_{i,k_0+l+1|k_0+1}$ to $\mathcal{S}_j$, $J \in \mathcal{P}_{-i}$;*

Step 2: Communicating at time k, $k > k_0$.

- *Measure $\mathbf{x}_{i,k}$, transmit $\mathbf{x}_{i,k}$, $\hat{\mathbf{x}}_{i,k+l+1|k}$, to $\mathcal{S}_j$, $j \in \mathcal{P}_{-i}$, and receive $\mathbf{x}_{j,k}$, $\hat{\mathbf{x}}_{j,k}$ from $\mathcal{S}_j$, $j \in \mathcal{P}_{+i}$;*

Step 3: Update control law at time k.

- *If $\mathbf{x}_k \in \Omega(\varepsilon)$, then apply the terminal controller $\mathbf{u}_{i,k} = \mathbf{K}_i\mathbf{x}_{i,k}$; Else*
- *Solve Problem 1. for $\mathbf{u}_{i,k+l-1|k}$ and apply $\mathbf{u}_{i,k|k}$ to $\mathcal{S}_i$;*
- *Compute $\hat{\mathbf{x}}_{i,k+l+1|k+1}$ according to (12) and transmit $\hat{\mathbf{x}}_{i,k+l+1|k+1}$ to $\mathcal{S}_j$, $J \in \mathcal{P}_{-i}$;*

Step 4: Update control at time $k+1$.

- *Let $k+1 \rightarrow k$, repeat Step 2.*

Algorithm 1. presumes that all local controllers $\mathcal{C}_i$, $i \in \mathcal{P}$, have access to the full state $\mathbf{x}_k$. This requirement results solely from the use of the dual-mode control, in which the switching occurs synchronously only when $\mathbf{x}_k \in \Omega(\varepsilon)$, with $\Omega(\varepsilon)$ being as defined in Lemma 3.1.. In the next section, it is will be shown that the CF-DMPC policy drives the state $\mathbf{x}_{k+l}$ to $\Omega(\varepsilon)$ in a finite number of updates. As a result, if $\Omega_i(\varepsilon)$ is chosen sufficiently small, then MPC can be employed for all time without switching to a terminal controller, eliminating the need of the local controllers to access the full state. Of course, in this case, instead of asymptotic stability at the origin, we can only drive the state toward the small set $\Omega(\varepsilon)$.

The analysis in the next section shows that the CF-DMPC algorithm is feasible at every update and is stabilizing.

4. Analysis

In this section, feasibility is analyzed in the first subsection, followed by stability analysis in the second subsection.

4.1. Feasibility

The main result of this section is that, provided that an initially feasible solution is available and Assumption 3. holds true, for any $\mathcal{S}_i$ and at any time $k \geq 1$, $\mathbf{u}^{\mathrm{p}}_{i,\cdot|k} = \mathbf{u}^{\mathrm{f}}_{i,\cdot|k}$ is a feasible control solution to Problem 1..

To establish this feasibility result, we will show that, for any $\mathcal{S}_i$ and at any update $k \geq 1$, the control and state pair $(\mathbf{u}^{\mathrm{f}}_{i,\cdot|k}, \mathbf{x}^{\mathrm{f}}_{i,\cdot|k})$ are a feasible solution to Problem 1. which satisfy the consistency constraints (17),(18), the control constraint (20) and the terminal state constraint (21).

Fig. 2 shows the discrepancy between the presumed state sequence $\{\hat{\mathbf{x}}_{ik+1|k}, \hat{\mathbf{x}}_{i,k+2|k}, \cdots\}$ and the predicted state sequence $\{\mathbf{x}^{\mathrm{f}}_{i,k+1|k}, \mathbf{x}^{\mathrm{f}}_{i,k+2|k}, \cdots\}$, $j \in \mathcal{P}_i$, and the relationships between these sequences and the terminal sets $\Omega_j(\varepsilon)$, $\Omega_j(\varepsilon/2)$ and $\Omega_j(\varepsilon'/2)$, where $0 < \varepsilon' = (1-\kappa)\varepsilon < \varepsilon$. To ensure feasibility, parametric conditions must be established under which $\hat{\mathbf{x}}_{i,k+N|k}$ and $\mathbf{x}^{\mathrm{f}}_{i,k+N|k}$ are within the indicated ellipsoids at the time instant k and $\hat{\mathbf{x}}_{i,k+s|k}$ and $\mathbf{x}^{\mathrm{f}}_{i,k+s|k}$ are sufficiently close to each other over the entire time interval $[k+1, k+N]$.

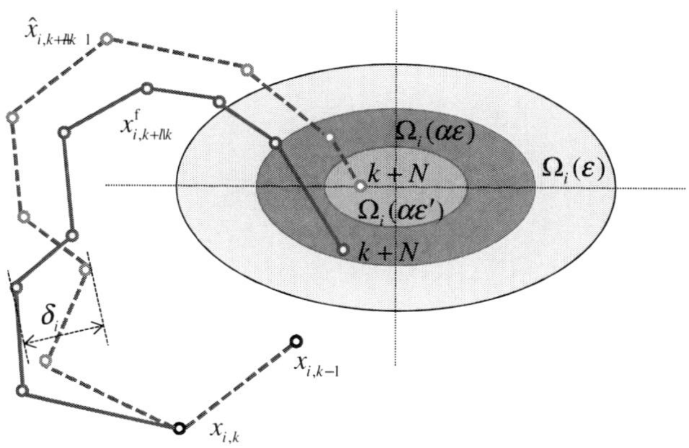

Figure 2. Schematic of the discrepancy among a feasible state sequence, the presumed state sequence and the predictive sequence.

Lemma 4.1. identifies sufficient conditions that ensure $\hat{x}_{i,k+N|k} \in \Omega_i(\varepsilon'/2)$, where $\varepsilon' = (1-\kappa)\varepsilon$. Lemma 4.2. identifies sufficient conditions that ensure $\|\mathbf{x}^{\mathrm{f}}_{i,l+k|k} - \hat{\mathbf{x}}_{i,s+k|k}\|_{\mathbf{P}_j} \leq$

$\kappa\varepsilon/(2\sqrt{m})$ for all $i \in \mathcal{P}$. Lemma 4.3. establishes that the control constraint is satisfied. Finally, Theorem 4.1. combines the results in Lemmas 4.1.-4.3. to arrive at the conclusion that, for any $i \in \mathcal{P}$, the control and state pair $(\mathbf{u}^{\mathrm{f}}_{i,\cdot|k}, \mathbf{x}^{\mathrm{f}}_{i,\cdot|k})$ are a feasible solution to Problem 1. at any update $k \geq 1$.

Lemma 4.1. *Suppose that Assumptions 1.-3. hold and* $\mathbf{x}_{k_0} \in \mathcal{X}$. *For any* $k \geq 0$, *if Problem 1. has a solution at time* $k-1$ *and* $\hat{\mathbf{x}}_{i,k+N-1|k-1} \in \Omega_j(\varepsilon/2)$ *for any* $j \in \mathcal{P}_i, i \in \mathcal{P}$, *then*

$$\hat{\mathbf{x}}_{i,k+N-1|k} \in \Omega_j(\varepsilon/2)$$

and

$$\hat{\mathbf{x}}_{i,k+N|k} \in \Omega_j(\varepsilon'/2),$$

provided that $\widehat{Q}_j$ *and* P_j *satisfy*

$$\max_{i\in\mathcal{P}}(\rho_i) \leq 1-\kappa \tag{25}$$

and $\varepsilon' = (1-\kappa)\varepsilon$,

$$\rho = \lambda_{\max}\sqrt{\left(\hat{\mathbf{Q}}_i\mathbf{P}_i^{-1}\right)^{\mathrm{T}}\hat{\mathbf{Q}}_i\mathbf{P}_i^{-1}}$$

Proof. Since Problem 1. has a solution at time $k-1$, by construction (13) and (14), it has

$$\left\|\hat{\mathbf{x}}_{i,k+N-1|k}\right\|_{\mathbf{P}_j} = \left\|\mathbf{x}^{\mathrm{p}}_{i,k+N-1|k-1}\right\|_{\mathbf{P}_i} \leq \frac{\varepsilon}{2\sqrt{m}} \tag{26}$$

In addition, since

$$\begin{aligned}\hat{\mathbf{x}}_{i,k+N|k} &= \mathbf{A}_{\mathrm{d}i}\mathbf{x}^{\mathrm{p}}_{i,k+N-1|k-1} + \sum_{j\in\mathcal{P}_{+i}}\mathbf{A}_{ij}\mathbf{x}^{\mathrm{p}}_{j,k+N-1|k-1}\\ &= \mathbf{A}_{\mathrm{d}i}\hat{\mathbf{x}}_{i,k+N-1|k} + \sum_{j\in\mathcal{P}_{+i}}\mathbf{A}_{ij}\hat{\mathbf{x}}_{j,k+N-1|k}\end{aligned} \tag{27}$$

It has

$$\left\|\hat{\mathbf{x}}_{i,k+N|k}\right\|_{\mathbf{P}_i} = \left\|\mathbf{A}_{\mathrm{d}i}\hat{\mathbf{x}}_{i,k+N-1|k} + \sum_{j\in\mathcal{P}_{+i}}\mathbf{A}_{ij}\hat{\mathbf{x}}_{j,k+N-1|k}\right\|_{\mathbf{P}_i} \tag{28}$$

Consider Assumption 2., $\mathbf{A}_{\mathrm{o}}^{\mathrm{T}}\mathbf{P}\mathbf{A}_{\mathrm{o}} + \mathbf{A}_{\mathrm{o}}^{\mathrm{T}}\mathbf{P}\mathbf{A}_{\mathrm{d}} + \mathbf{A}_{\mathrm{d}}^{\mathrm{T}}\mathbf{P}\mathbf{A}_{\mathrm{o}} < \hat{\mathbf{Q}}/2$
Thus if , then

$$\begin{aligned}\left\|\hat{\mathbf{x}}_{i,k+N|k}\right\|_{\mathbf{P}_i} &\leq \left\|\hat{\mathbf{x}}_{i,k+N-1|k}\right\|_{\hat{\mathbf{Q}}/2}\\ &\leq \lambda_{\max}\sqrt{\left(\hat{\mathbf{Q}}_i\mathbf{P}_i^{-1}\right)^{\mathrm{T}}\hat{\mathbf{Q}}_i\mathbf{P}_i^{-1}}\left\|\hat{\mathbf{x}}_{i,k+N-1|k}\right\|_{\mathbf{P}_i}\\ &\leq (1-\kappa)\frac{\varepsilon}{2\sqrt{m}}\end{aligned} \tag{29}$$

This completes the proof of Lemma 4.1. □

Lemma 4.2. *Suppose that Assumptions 1.-3. hold and $x(k_0) \in \mathcal{X}$. For any $k \geq 0$, if Problem 1. has a solution at every update time l, $l = 1, 2, \cdots, k-1$, then*

$$\left\| \mathbf{x}^{\mathrm{f}}_{i,k+l|k} - \hat{\mathbf{x}}_{i,k+l|k} \right\|_{\mathbf{P}_i} \leq \frac{\kappa\varepsilon}{2\sqrt{m}}, \tag{30}$$

for all $i \in \mathcal{P}_i$ and all $l = 1, 2, \cdots, N$, provided that (25) and the following parametric condition hold

$$\frac{\sqrt{m_2}}{\xi\lambda_{\min}(P)} \sum_{l=0}^{N-2} \alpha_l \leq 1, \tag{31}$$

where α_l is as defined in (23). Furthermore, the feasible control $\mathbf{u}^{\mathrm{f}}_{i,k+s|k}$ and the feasible state $\mathbf{x}^{\mathrm{f}}_{i,k+l|k}$ satisfy constraints (17)-(18).

Proof. We will prove (30) first. Since a solution exists at update time $1, 2, \ldots, k-1$, according to (12), (13) and (24), for any $s = 1, 2, \cdots, N-1$, the feasible state is given by

$$\begin{aligned} \mathbf{x}^{\mathrm{f}}_{i,k+l|k} &= \mathbf{A}^{l}_{ii}\mathbf{x}^{\mathrm{f}}_{i,k|k} + \sum_{h=1}^{l} \mathbf{A}^{l-h}_{ii}\mathbf{B}_{ii}\mathbf{u}^{\mathrm{f}}_{i,k+l|k} + \sum_{j\in\mathcal{P}_{+i}} \sum_{h=1}^{l} \mathbf{A}^{l-h}_{ii}\mathbf{A}_{ij}\hat{\mathbf{x}}_{j,k+h-1|k} \\ &= \mathbf{A}^{l}_{ii}\left(\mathbf{A}^{l}_{ii}\mathbf{x}_{i,k-1|k-1} + \mathbf{B}_{ii}\mathbf{u}_{i,k-1|k-1} + \sum_{j\in\mathcal{P}_{+i}} \mathbf{A}_{ij}\mathbf{x}_{j,k-1|k-1}\right) \\ &\quad + \sum_{h=1}^{l} \mathbf{A}^{l-h}_{ii}\mathbf{B}_{ii}\hat{\mathbf{u}}_{i,k+l|k} + \sum_{j\in\mathcal{P}_{+i}} \sum_{h=1}^{l} \mathbf{A}^{l-h}_{ii}\mathbf{A}_{ij}\mathbf{x}^{\mathrm{p}}_{j,k+h-1|k-1} \end{aligned} \tag{32}$$

and the presumed state is

$$\begin{aligned} \hat{\mathbf{x}}_{i,k+l|k} &= \mathbf{A}^{l}_{ii}\mathbf{x}_{i,k|k-1} + \sum_{h=1}^{l} \mathbf{A}^{l-h}_{ii}\mathbf{B}_{ii}\mathbf{u}_{i,k+l|k-1} + \sum_{j\in\mathcal{P}_{+i}} \sum_{h=1}^{l} \mathbf{A}^{l-h}_{ii}\mathbf{A}_{ij}\hat{\mathbf{x}}_{j,k+h-1|k-1} \\ &= \mathbf{A}^{l}_{ii}\left(\mathbf{A}^{l}_{ii}\mathbf{x}_{i,k-1|k-1} + \mathbf{B}_{ii}\mathbf{u}_{i,k-1|k-1} + \sum_{j\in\mathcal{P}_{+i}} \mathbf{A}_{ij}\hat{\mathbf{x}}_{j,k-1|k-1}\right) \\ &\quad + \sum_{h=1}^{l} \mathbf{A}^{l-h}_{ii}\mathbf{B}_{ii}\hat{\mathbf{u}}_{i,k+l|k} + \sum_{j\in\mathcal{P}_{+i}} \sum_{h=1}^{l} \mathbf{A}^{l-h}_{ii}\mathbf{A}_{ij}\hat{\mathbf{x}}_{j,k+h-1|k-1} \end{aligned} \tag{33}$$

Subtracting (34) from (33), and from the definition of (23) we obtain the discrepancy between the feasible state sequence and the presumed state sequence as

$$\begin{aligned} \left\| \mathbf{x}^{\mathrm{f}}_{i,k+l|k} - \hat{\mathbf{x}}_{i,k+l|k} \right\|_{\mathbf{P}_i} &= \left\| \sum_{j\in\mathcal{P}_{+i}} \sum_{h=1}^{l} \mathbf{A}^{l-h}_{ii}\mathbf{A}_{ij}\left(\mathbf{x}^{\mathrm{p}}_{i,k+h-1|k-1} - \hat{\mathbf{x}}_{j,k+h-1|k-1}\right) \right\|_{\mathbf{P}_i} \\ &\leq \sum_{j\in\mathcal{P}_{+i}} \sum_{h=1}^{l} \mathbf{A}^{l-h}_{ii}\mathbf{A}_{ij} \left\| \mathbf{x}^{\mathrm{p}}_{i,k+h-1|k-1} - \hat{\mathbf{x}}_{j,k+h-1|k-1} \right\|_{\mathbf{P}_i} \\ &\leq \sum_{s=1}^{l} \alpha_{l-s} \left\| \mathbf{x}^{\mathrm{p}}_{i,k+s-1|k-1} - \hat{\mathbf{x}}_{i,k+s-1|k-1} \right\|_{2}. \end{aligned} \tag{34}$$

Let $\mathcal{S}_r$ be the subsystem which maximizes

$$\sum_{h=1}^{l} \alpha_{l-h} \left\| \mathbf{x}^{\mathrm{p}}_{i,k-1+h|k-1} - \hat{\mathbf{x}}_{i,k-1+h|k-1} \right\|_2, i \in \mathcal{P} \tag{35}$$

Then, the following equation can be deduced from (35)

$$\left\| \mathbf{x}^{f}_{j,k+l|k} - \hat{\mathbf{x}}_{j,k+l|k} \right\|_{\mathbf{P}_i} \leq \sqrt{m_1} \sum_{h=1}^{l} \alpha_{l-h} \left\| \mathbf{x}^{\mathrm{p}}_{g,k+h-1|k-1} - \hat{\mathbf{x}}_{g,k+h-1|k-1} \right\|_2 \tag{36}$$

Since $\mathbf{x}^{\mathrm{p}}_{i,l|k-1}$ satisfy constraints (17) for all times $l = 1, 2, \ldots, k-1$, the following equation can be deduced

$$\begin{aligned} & \left\| \mathbf{x}^{\mathrm{f}}_{i,k+l|k} - \hat{\mathbf{x}}_{i,k+l|k} \right\|_{\mathbf{P}_i} \\ \leq\ & \frac{(1-\xi)(1-\kappa)\varepsilon}{2\sqrt{m}} + \frac{\xi(1-\kappa)\varepsilon}{2\sqrt{m}} \\ =\ & \frac{\kappa\varepsilon}{2\sqrt{m}}. \end{aligned} \tag{37}$$

Thus, (30) holds for all $l = 1, 2, \ldots, N-1$.

When $l = N$, it has

$$\mathbf{x}^{\mathrm{f}}_{i,k+N|k} = \mathbf{A}_{\mathrm{d},i} \mathbf{x}^{\mathrm{f}}_{i,k+N|k} + \sum_{j \in \mathcal{P}_{+i}} \mathbf{A}_{ij} \hat{\mathbf{x}}_{j,k+N-1|k} \tag{38}$$

$$\hat{\mathbf{x}}_{i,k+N|k} = \mathbf{A}_{\mathrm{d},i} \hat{\mathbf{x}}_{i,k+N-1|k} + \sum_{j \in \mathcal{P}_{+i}} \mathbf{A}_{ij} \hat{\mathbf{x}}_{j,k+N-1|k} P \tag{39}$$

From the subtraction of the two equations, then, the discrepancy between the feasible state $\mathbf{x}^{\mathrm{f}}_{i,(k+N)}$ and the presumed state $\hat{\mathbf{x}}_{i,(k+N)}$ is

$$\mathbf{x}^{\mathrm{f}}_{i,k+N|k} - \hat{\mathbf{x}}_{i,k+N|k} = \mathbf{A}_{\mathrm{d},i} \left(\mathbf{x}^{\mathrm{f}}_{i,k+N-1|k} - \hat{\mathbf{x}}_{i,k+N-1|k} \right) \tag{40}$$

Consequently, (30) holds for all $l = 1, 2, \ldots, N$.

In what follows we will prove that the feasible state $\mathbf{x}^{\mathrm{f}}_{i,(k+l)}$ satisfy constraints (17)(18) when (30) holds.

When $l = 1, 2, \ldots, N-1$, substitute $\mathbf{x}^{\mathrm{f}}_{i,k+l|k}$ in the constraint (17) with considering (31), we can get

$$\begin{aligned} & \sum_{h=1}^{l} \alpha_{l-h} \left\| \mathbf{x}^{\mathrm{f}}_{i,k+h|k} - \hat{\mathbf{x}}_{i,k+h|k} \right\|_2 \\ \leq\ & \frac{1}{\lambda_{\min}(\mathbf{P}_i)} \sum_{l=1}^{s} \alpha_{l-h} \left\| \mathbf{x}^{\mathrm{f}}_{i,k+h|k} - \hat{\mathbf{x}}_{i,k+h|k} \right\|_{\mathbf{P}_i} \\ \leq\ & \frac{1}{\lambda_{\min}(\mathbf{P})} \sum_{h=1}^{l} \alpha_{l-h} \frac{\sqrt{m_1}}{\xi} \frac{\xi\kappa\varepsilon}{2\sqrt{mm_1}} \end{aligned} \tag{41}$$

Thus, when

$$\frac{\sqrt{m_1}}{\xi\lambda_{\min}(\mathbf{P})}\sum_{h=1}^{l}\alpha_{l-h}\leq 1,$$

state $\mathbf{x}^{\mathrm{f}}_{i,k+l|k}$, $l=1,2,\ldots,N-1$, satisfy constraint constraint (17).

Finally, when $l=N$,

$$\left\|\mathbf{x}^{\mathrm{f}}_{i,k+N|k}-\hat{\mathbf{x}}_{i,k+N|k}\right\|_{\mathbf{P}_i}\leq\frac{\kappa\varepsilon}{2\sqrt{m}},\tag{42}$$

which shows constraint (18) is satisfied. The proof is completed. □

In what follows we establish that, at time k, if the conditions (25),(30) and (31) are satisfied, then $\mathbf{u}^{\mathrm{f}}_{i,k+l-1|k}$, $l=1,2,\ldots,N$, is feasible solution of Problem 1..

Lemma 4.3. *Suppose that Assumptions 1.-3. hold, $\mathbf{x}_{k_0}\in\mathbf{R}^{n_x}$, and conditions (25) and (31) are satisfied. For any $k\geq 0$, if Problem 1. has a solution at every update time t, $t=1,2,\cdots,k-1$, then $\mathbf{u}^{\mathrm{f}}_{i,k+l|k}\in\mathcal{U}$ for all $l=1,2,\cdots,N-1$.*

Proof. Since Problem 1. has a feasible solution at $t=1,2,\cdots,k-1$, and $\mathbf{u}^{\mathrm{f}}_{i,k+s-1|k}=\mathbf{u}^{\mathrm{p}}_{i,k+s-1|k-1}$ for all $l=1,2,\cdots,N-1$, we only need to show that $\mathbf{u}^{\mathrm{f}}_{i,k+N-1|k}\in\mathcal{U}$.

Since ε has been chosen to satisfy the conditions of Lemma 3.1., $\mathbf{K}_i\mathbf{x}_i\in\mathcal{U}$ for all $i\in\mathcal{P}$ when $\mathbf{x}\in\Omega(\varepsilon)$. Consequently, a sufficient condition for $\mathbf{u}^{\mathrm{f}}_{i,k+N-1|k}\in\mathcal{U}$ is that $\mathbf{x}^{\mathrm{f}}_{i,k+N-1|k}\in\Omega(\varepsilon)$.

In view of Lemmas 4.1. and 4.2., using the triangle inequality, we have

$$\begin{aligned}
&\|\mathbf{x}^{\mathrm{f}}_{i,k+N-1|k}\|_{\mathbf{P}_i}\\
\leq\;&\left\|\mathbf{x}^{\mathrm{f}}_{i,k+N-1|k}-\hat{\mathbf{x}}_{i,k+N-1|k}\right\|_{\mathbf{P}_i}\\
&+\left\|\hat{\mathbf{x}}_{i,k+N-1|k}\right\|_{\mathbf{P}_i}\\
\leq\;&\frac{\varepsilon}{2(q+1)\sqrt{m}}+\frac{\varepsilon}{2\sqrt{m}}\\
\leq\;&\frac{\varepsilon}{\sqrt{m}},
\end{aligned}$$

that is, $\mathbf{x}^{\mathrm{f}}_{i,k+N|k}\in\Omega_i(\varepsilon)$. This concludes the proof. □

Lemma 4.4. *Suppose that Assumptions 1. and 3. hold, $\mathbf{x}_{k_0}\in\mathcal{X}$, and the conditions (30) and (31) are satisfied. For any $k\geq 0$, if Problem 1. has a solution at every update time t, $t=0,\ldots,k-1$, then the terminal state constraint $\mathbf{x}^{\mathrm{f}}_{i,k+N|k}\in\Omega\left(\varepsilon/2\right)$ is hold, for any $i\in\mathcal{P}$.*

Proof. Since there is a solution for Problem 1. at updates $t=1,\ldots,k-1$, Lemma 4.1.-4.3.

can be invoked. Using the triangle inequality, it has

$$\begin{aligned}
& \|\mathbf{x}^{\mathrm{f}}_{i,k+N|k}\|_{\mathbf{P}_i} \\
\leq\ & \|\mathbf{x}^{\mathrm{f}}_{i,k+N|k} - \hat{\mathbf{x}}_{i,k+N|k-1}\|_{\mathbf{P}_i} + \|\hat{\mathbf{x}}_{i,k+N|k-1,i}\|_{\mathbf{P}_i} \\
\leq\ & \frac{\kappa\varepsilon}{2\sqrt{m}} + \frac{(1-\kappa)\varepsilon}{2\sqrt{m}} \\
=\ & \frac{\varepsilon}{2\sqrt{m}}
\end{aligned} \tag{43}$$

for each $i \in \mathcal{P}$. This shows that the terminal state constraint is satisfied. This completes the proof. □

Theorem 4.1. *Suppose that Assumptions 1.-3. hold,* $\mathbf{x}(k_0) \in \mathbf{R}^{n_x}$ *and constraints (17),(18) and (20) are satisfied at* k_0*. Then, for every* $i \in \mathcal{P}$*, the control* $\mathbf{u}^{\mathrm{f}}_{i,\cdot|k}$ *and state* $\mathbf{x}^{\mathrm{f}}_{i,\cdot|k}$*, respectively defined by (24), are a feasible solution of Problem 1. at every update* $k \geq 1$.

Proof. We will prove the theorem by induction.

First, consider the case of $k = 1$. The state sequence $\mathbf{x}^{\mathrm{p}}_{i,\cdot|1} = \mathbf{x}^{\mathrm{f}}_{i,\cdot|1}$ trivially satisfies the dynamic equation (12), the stability constraint (19) and the consistency constraints (17)-(18).

Observe that

$$\hat{\mathbf{x}}_{i,1|1} = \mathbf{x}^{\mathrm{p}}_{i,1|0} = \mathbf{x}^{\mathrm{f}}_{i,1|1} = \mathbf{x}_{i,1},\ i \in \mathcal{P},$$

and that

$$\mathbf{x}^{\mathrm{f}}_{i,1+l|1} = \mathbf{x}^{\mathrm{p}}_{i,1+l|0},\ l = 1, 2, \cdots, N-1.$$

Thus, $x^{\mathrm{f}}_{i,N|1} \in \Omega_i(\varepsilon/2)$. By the invariance of $\Omega(\varepsilon)$ under the terminal controller and the conditions in Lemma 3.1., it follows that the terminal state and control constraints are also satisfied. This completes the proof of the case of $k = 1$.

Now suppose that $\mathbf{u}^{\mathrm{p}}_{i,\cdot|l} = \mathbf{u}^{\mathrm{f}}_{i,\cdot|l}$ is a feasible solution for $l = 1, 2, \cdots, k-1$. We will show that $\mathbf{u}^{\mathrm{f}}_{i,\cdot|k}$ is a feasible solution at update k.

As before, the consistency constraint (17) is trivially satisfied, and $\mathbf{u}^{\mathrm{f}}_{i,\cdot|k}$ is the corresponding state sequence that satisfies the dynamic equation. Since there is a solution for Problem 1. at updates $l = 1, 2, \cdots, k-1$, Lemmas 4.1.-4.3. can be invoked. Lemma 4.3. guarantees control constraint feasibility. Lemma 4.4. shows that the terminal state constraint is satisfied and the proof of Theorem 4.1. is completed. □

4.2. Stability

The stability of the closed-loop system is analyzed in this subsection.

Theorem 4.2. *Suppose that Assumptions 1.-3. hold,* $x(k_0) \in \mathbf{R}^{n_x}$*, constraints (17)-(19) and (20) are satisfied, and the following parametric condition holds*

$$\kappa\frac{N-1}{2} + \frac{1}{\mu} < \frac{1}{2}. \tag{44}$$

Then, by application of Algorithm 1., the closed-loop system (2) is asymptotically stable at the origin.

Proof. By Algorithm 1. and Lemma 3.1., when $\mathbf{x}(k)$ enters $\Omega(\varepsilon)$, the terminal controllers take over to keep it in there and stabilize the system at the origin. Therefore, it remains to show that if $\mathbf{x}(0) \in \mathcal{X} \setminus \Omega(\varepsilon)$, then by the application of Algorithm 1., the state of system (2) is driven to the set $\Omega(\varepsilon)$ in finite time.

Define the non-negative function for $\mathcal{S}$

$$V_k = \sum\nolimits_{l=1}^{N} \|\mathbf{x}^{\mathrm{p}}_{k+l|k}\|_{\mathbf{P}}.$$

In what follows, we will show that, for any $k \geq 0$, if $\mathbf{x}_k \in \mathcal{X} \setminus \Omega(\varepsilon)$, then there exists a constant $\eta \in (0, \infty)$ such that $V_k \leq V_{k-1} - \eta$. Constraint (19) implies that

$$\left\|\mathbf{x}^{\mathrm{p}}_{k+l|k}\right\|_{\mathbf{P}} \leq \left\|\mathbf{x}^{\mathrm{f}}_{k+l|k}\right\|_{\mathbf{P}} + \frac{\varepsilon}{\mu N}.$$

Therefore,

$$V_k \leq \sum\nolimits_{l=1}^{N} \left\|\mathbf{x}^{\mathrm{f}}_{k+l|k}\right\|_{\mathbf{P}} + \frac{\varepsilon}{\mu}.$$

Subtracting V_{k-1} from V_k and using $\mathbf{x}^{\mathrm{p}}_{k+l|k-1} = \hat{\mathbf{x}}_{k+l|k}$, $l = 1, 2, \cdots, N-1$, gives

$$\begin{aligned} & V_k - V_{k-1} \\ \leq \; & -\left\|\mathbf{x}^{\mathrm{p}}_{k|k-1}\right\|_{\mathbf{P}} + \frac{\varepsilon}{\mu} + \left\|\mathbf{x}^{\mathrm{f}}_{k+N|k}\right\|_{\mathbf{P}} \\ & + \sum_{l=1}^{N-1} \left(\left\|\mathbf{x}^{\mathrm{f}}_{k+l|k}\right\|_{\mathbf{P}} - \|\hat{\mathbf{x}}_{k+l|k}\|_{\mathbf{P}} \right). \end{aligned} \tag{45}$$

Assuming $\mathbf{x}_k \in \mathcal{X} \setminus \Omega(\varepsilon)$ yields

$$\left\|\mathbf{x}^{\mathrm{p}}_{k|k-1}\right\|_{\mathbf{P}} > \varepsilon. \tag{46}$$

Also, by Theorem 4.1. we have

$$\left\|\mathbf{x}^{\mathrm{f}}_{k+N|k)}\right\|_{\mathbf{P}} \leq \varepsilon/2, \tag{47}$$

and by Lemma 4.2., we have

$$\sum_{l=1}^{N-1} \left(\left\|\mathbf{x}^{\mathrm{f}}(k+l|k)\right\|_{\mathbf{P}} - \|\hat{\mathbf{x}}_{k+l|k}\|_{\mathbf{P}} \right) \leq \frac{(N-1)\kappa\varepsilon}{2}. \tag{48}$$

Using (46)-(48) in (45) then yields

$$V_k - V_{k-1} < \varepsilon \left(-1 + \frac{(N-1)\kappa}{2} + \frac{1}{2} + \frac{1}{\mu} \right), \tag{49}$$

which, in view of (44), implies that $V_k - V_{k-1} < 0$. Thus, for any $k \geq 0$, if $\mathbf{x}_k \in \mathcal{X} \setminus \Omega(\varepsilon)$, there is a constant $\eta \in (0, \infty)$ such that $V_k \leq V_{k-1} - \eta$. It then follows that there exists a finite time k' such that $\mathbf{x}_{k'} \in \Omega(\varepsilon)$. This concludes the proof. □

We have now established the feasibility the CF-DMPC and the stability of the resulting closed-loop system. That is, if an initially feasible solution could be found, subsequent feasibility of the algorithm is guaranteed at every update, and the resulting closed-loop system is asymptotically stable at the origin.

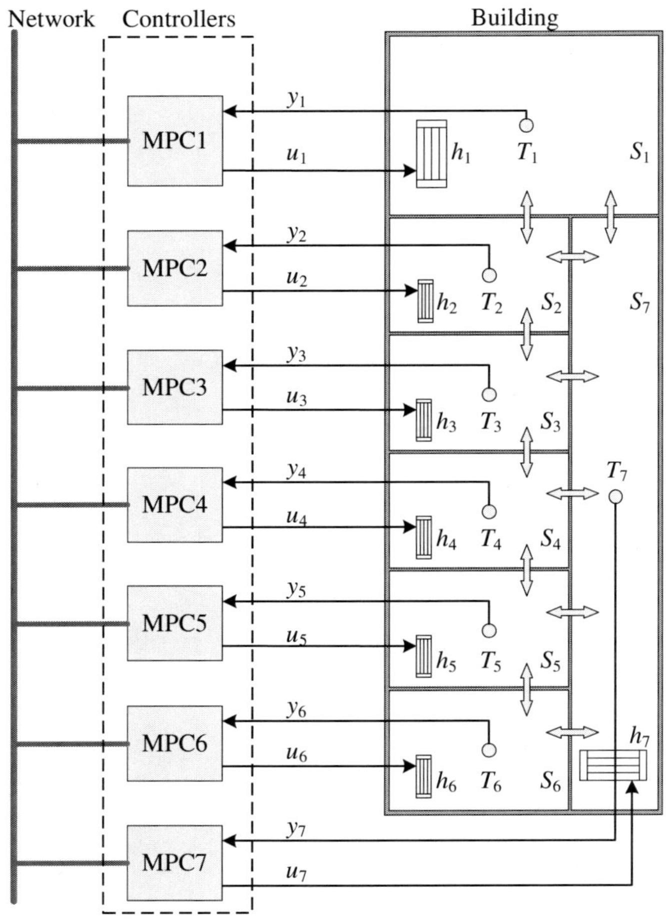

Figure 3. The interaction relationship among subsystems.

5. Simulation

The multi-zones building temperature regulation systems are a class of typical spatially distributed systems, as shown in Fig. 3, which are composed of many physically interacted subsystems (rooms or zones) labeled with $\mathcal{S}_1, \mathcal{S}_2, \ldots$, respectively. The thermal influences between rooms of the same building occur through internal walls (the internal walls isolation is weak) and/or door openings. A thermal-meter and a heater (or air condition) are installed in each zone, which is used to measure and adjust the temperature of the multi-zones building.

For simplicity, the 7-zones building is taken as example. The relationship among these seven zones is also shown in Fig. 3., where zone $\mathcal{S}_1$ is impacted by zone $\mathcal{S}_2$ and zone $\mathcal{S}_7$; zone $\mathcal{S}_2$ is impacted by zone $\mathcal{S}_1$, $\mathcal{S}_3$ and zone $\mathcal{S}_7$; zone $\mathcal{S}_3$ is impacted by zone $\mathcal{S}_2$, $\mathcal{S}_4$ and zone $\mathcal{S}_7$; zone $\mathcal{S}_4$ is impacted by zone $\mathcal{S}_3$, $\mathcal{S}_5$ and zone $\mathcal{S}_7$; zone $\mathcal{S}_5$ is impacted by zone $\mathcal{S}_4$, $\mathcal{S}_6$ and zone $\mathcal{S}_7$; zone $\mathcal{S}_6$ is impacted by zone $\mathcal{S}_5$ and zone $\mathcal{S}_7$; zone $\mathcal{S}_4$ is impacted by all the other zones.

Let $\mathcal{U}_i$ be defined to reflect both the constraint on the input $u_i \in [u_{i,\mathrm{L}}, u_{i,\mathrm{U}}]$ and the constraint on the increment of the input $\Delta u_i \in [\Delta u_{i,\mathrm{L}}, \Delta u_{i,\mathrm{U}}]$.

The models of these seven subsystems are respectively given by

$$\begin{aligned}
x_{1,k+1} &= 0.574x_{1,k} + 0.384u_{1,k} + 0.029x_{2,k} + 0.057x_{7,k} \\
x_{2,k+1} &= 0.535x_{2,k} + 0.372u_{2,k} + 0.054x_{1,k} + 0.054x_{3,k} \\
&\quad +0.054x_{7,k} \\
x_{3,k+1} &= 0.547x_{3,k} + 0.376u_{3,k} \\
&\quad +0.055x_{2,k} + 0.055x_{4,k} + 0.055x_{7,k} \\
x_{4,k+1} &= 0.606x_{4,k} + 0.394u_{4,k} \\
&\quad +0.061x_{3,k} + 0.061x_{5,k} + 0.061x_{7,k}, \\
x_{5,k+1} &= 0.681x_{5,k} + 0.415u_{5,k} \\
&\quad +0.068x_{4,k} + 0.068x_{6,k} + 0.068x_{7,k} \\
x_{6,k+1} &= 0.548x_{6,k} + 0.376u_{6,k} \\
&\quad +0.055x_{5,k} + 0.055x_{7,k}, \\
x_{7,k+1} &= 0.716x_{7,k} + 0.425u_{7,k} + 0.018x_{1,k} + 0.018x_{2,k} \\
&\quad +0.018x_{3,k} + 0.018x_{4,k} + 0.018x_{5,k} + 0.018x_{6,k}.
\end{aligned}$$

For the purpose of comparison, the centralized MPC, the DMPC where each subsystem-based MPC optimizes its own local cost [10], here we call it as local cost optimization based DMPC (LCO-DMPC), and the propsed CF-DMPC are all applied to this system.

Some parameters of the controllers in proposed CF-DMPC are shown in Table 2. Among these parameters, P_i is obtained by solving the Lyapnov function. The eigenvalue of each closed-loop system under the feedback control shown in the table is 0.5. Set $\varepsilon = 0.15$, and set the control horizon of all the controllers to be $N = 10$. Set the initial presumed inputs and states, at time $k_0 = 0$, be zeros.

In both the centralized MPC and the subsystem-based MPCs of the local cost optimization based DMPC, the dual mode strategy is adopted, and set the parameters, the initial states and the initial presumed inputs be the same as those used in CF-DMPC.

The state responses and the inputs of the closed-loop system under the control of the centralized MPC, CF-DMPC and LCO-DMPC are shown in Figs. 4. The shape of the state response curves under the control CF-DMPC are similar to those under the centralized MPC. Under the CF-DMPC control design, when set point changed, there is no significant overshooting, but some fluctuations exist in the trajectories of states of the interacting subsystems. Under the LCO-DMPC control design, the states of all subsystems could converge

Table 2. Parameters of CF-DMPC

$Sub-$ $system$	K_i	P_i	Q_i	R_i	$\Delta u_{i,\mathrm{U}}$ $\Delta u_{i,\mathrm{L}}$
$\mathcal{S}_1$	-0.44	5.38	4	0.2	± 1
$\mathcal{S}_2$	-0.34	5.36	4	0.2	± 1
$\mathcal{S}_3$	-0.37	5.37	4	0.2	± 1
$\mathcal{S}_4$	-0.52	5.40	4	0.2	± 1
$\mathcal{S}_5$	-0.68	5.46	4	0.2	± 1
$\mathcal{S}_6$	-0.37	5.37	4	0.2	± 1
$\mathcal{S}_7$	-0.76	5.49	4	0.2	± 1

Table 3. State square errors of the closed-loop system under the control of the centralized MPC (CMPC), the LCO-DMPC and the CF-DMPC

Items	CMPC	CF-DMPC	LCO-DMPC
$\mathcal{S}_1$	0.0109	0.1146	2.0891
$\mathcal{S}_2$	2.2038	3.0245	6.2892
$\mathcal{S}_3$	5.4350	6.9908	10.6391
$\mathcal{S}_4$	2.2480	3.2122	15.3015
$\mathcal{S}_5$	4.5307	5.6741	30.2392
$\mathcal{S}_6$	4.3403	5.4926	8.2768
$\mathcal{S}_7$	9.2132	11.0574	33.6902
Total	27.9819	35.5663	106.5251

to set point, but there exists much larger overshooting comparing to those under the control of FC-DMPC and centralized MPC, and there are larger amplitude in the fluctuating of state than those under the control of CF-DMPC.

Table 3 shows the state square errors of the closed-loop system under the control of the centralized MPC, the CF-DMPC and the local cost optimization based DMPC, respectively. The total errors under the CF-DMPC is 7.5844 (27.1%) larger than that under the centralized MPC. The total errors resulting from the LCO-DMPC is 78.5432 (280.7%) larger than that results from the centralized MPC. The performance of the CF-DMPC is significantly better than that of the LCO-DMPC.

Table 4 shows the required network connectivity under the control of the centralized MPC, the CF-DMPC and the LCO-DMPC, respectively. The required network connectivity under the control of CF-DMPC equals to that under the control of LCO-DMPC and is much less than that under the control of centralized MPC.

From these simulation results, it can be seen that the proposed CF-DMPC is able to steer the system states to the set point if there is a feasible solution at the initial states, and it could obtain a better global performance than LCO-DMPC when the same network connectivity provided. The global performance of entire closed-loop system is improved without any weakening of the characteristics of good error tolerance and high flexibility of the whole control system.

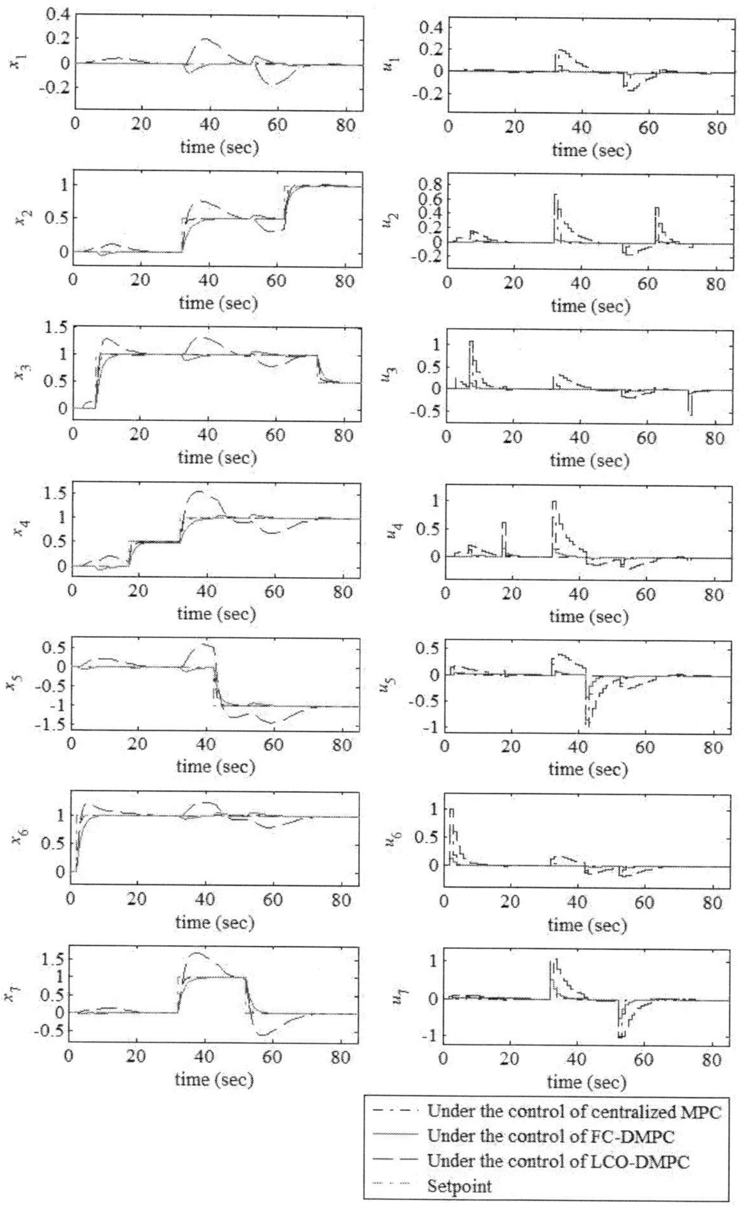

Figure 4. The evolution of the states under the centralized MPC, LCO-DMPC and CF-DMPC.

6. Conclusion

In this paper, a Coordinated Flexible DMPC proposed for distributed system is developed for dynamically coupled spatially distributed systems subject to decoupled input constraints. The proposed method could improve the global performance of entire closed-loop system without any increasing of network connectivity. In addition, if an initially feasible solution and a feed back control law $K_i x$ could be found, the subsequent feasibility of the algorithm is guaranteed at every update, and the resulting closed-loop system is asymptot-

Table 4. Required network connectivity under the control of the centralized MPC (CMPC), the LCO-DMPC and the CF-DMPC

Items	CMPC	CF-DMPC	LCO-DMPC
$\mathcal{S}_1$	All	2,7	2,7
$\mathcal{S}_2$	All	1,3,7	1, 3,7
$\mathcal{S}_3$	All	2, 4,7	2, 4,7
$\mathcal{S}_4$	All	3,5,7	3,5,7
$\mathcal{S}_5$	All	4,6,7	4,6,7
$\mathcal{S}_6$	All	5,7	5,7
$\mathcal{S}_7$	All	All	All

ically stable. The simulations illustrate that the performance of global system under the control of proposed method is very close to that under the control of centralized MPC.

Acknowledgment

This work was supported by the National Nature Science Foundation of China (61233004, 61221003, 61374109, 61304078,61473187), the National Basic Research Program of China (973 Program-2013CB035500), and partly sponsored by the International Cooperation Program of Shanghai Science and Technology Commission (12230709600), the Higher Education Research Fund for the Doctoral Program of China (20120073130006 & 20110073110018), and the China Postdoctoral Science Foundation (2013M540364).

References

[1] Walid Al-Gherwi, Hector Budman, and Ali Elkamel. Selection of control structure for distributed model predictive control in the presence of model errors. *Journal of Process Control*, 20(3):270–284, 2010.

[2] Walid Al-Gherwi, Hector Budman, and Ali Elkamel. A robust distributed model predictive control algorithm. *Journal of Process Control*, 21(8):1127–1137, 2011.

[3] I Alvarado, D Limon, D Muñoz de la Peña, JM Maestre, MA Ridao, H Scheu, W Marquardt, RR Negenborn, B De Schutter, F Valencia, et al. A comparative analysis of distributed mpc techniques applied to the hd-mpc four-tank benchmark. *Journal of Process Control*, 21(5):800–815, 2011.

[4] E. Camponogara, D. Jia, B.H. Krogh, and S. Talukdar. Distributed model predictive control. *IEEE Control Systems Magazine*, 22(1):44–52, 2002.

[5] Eduardo Camponogara and Helton F Scherer. Distributed optimization for model predictive control of linear dynamic networks with control-input and output constraints. *IEEE Transactions on Automation Science and Engineering*, 8(1):233–242, 2011.

[6] Ruoyu Cheng, J Fraser Forbes, and W San Yip. Dantzig–wolfe decomposition and plant-wide mpc coordination. *Computers & Chemical Engineering*, 32(7):1507–1522, 2008.

[7] Panagiotis D Christofides, Riccardo Scattolini, David Muñoz de la Peña, and Jinfeng Liu. Distributed model predictive control: A tutorial review and future research directions. *Computers & Chemical Engineering*, 51:21–41, 2013.

[8] Alejandro J del Real, Alicia Arce, and Carlos Bordons. Combined environmental and economic dispatch of smart grids using distributed model predictive control. *International Journal of Electrical Power & Energy Systems*, 54:65–76, 2014.

[9] X. Du, Y. Xi, and S. Li. Distributed model predictive control for large-scale systems. In *American Control Conference, 2001. Proceedings of the 2001*, volume 4, pages 3142–3143. IEEE, 2001.

[10] W.B. Dunbar. Distributed receding horizon control of dynamically coupled nonlinear systems. *IEEE Transactions on Automatic Control*, 52(7):1249–1263, 2007.

[11] Marcello Farina and Riccardo Scattolini. Distributed predictive control: a non-cooperative algorithm with neighbor-to-neighbor communication for linear systems. *Automatica*, 48(6):1088–1096, 2012.

[12] V Kirubakaran, TK Radhakrishnan, and N Sivakumaran. Distributed multiparametric model predictive control design for a quadruple tank process. *Measurement*, 47:841–854, 2014.

[13] J.W. Lee, W. Hyun Kwon, and J. Choi. On stability of constrained receding horizon control with finite terminal weighting matrix. *Automatica*, 34(12):1607–1612, 1998.

[14] S. Leirens, C. Zamora, RR Negenborn, and B. De Schutter. Coordination in urban water supply networks using distributed model predictive control. In *American Control Conference (ACC), 2010*, pages 3957–3962. IEEE, 2010.

[15] S. Li, Y. Zhang, and Q. Zhu. Nash-optimization enhanced distributed model predictive control applied to the shell benchmark problem. *Information Sciences*, 170(2-4):329–349, 2005.

[16] S. Li, Y. Zheng, and Z. Lin. Impacted-region optimization for distributed model predictive control systems with constraints. *Automation Science and Engineering, IEEE Transactions on*, PP(99):1–14, 2014.

[17] S. Li, Y. Zheng, and B. Wang. Steady-state target calculation for constrained predictive control systems based on goal programming. *Asia-Pacific Journal of Chemical Engineering*, 3(6):648–655, 2008.

[18] Jinfeng Liu, David Muñoz de la Peña, and Panagiotis D Christofides. Distributed model predictive control of nonlinear systems subject to asynchronous and delayed measurements. *Automatica*, 46(1):52–61, 2010.

[19] J.M. Maciejowski. *Predictive control: with constraints*. Pearson education, 2002.

[20] D.Q. Mayne, J.B. Rawlings, C.V. Rao, and P.O.M. Scokaert. Constrained model predictive control: Stability and optimality. *Automatica*, 36(6):789–814, 2000.

[21] Paul Mc Namara, Rudy R Negenborn, Bart De Schutter, and Gordon Lightbody. Weight optimisation for iterative distributed model predictive control applied to power networks. *Engineering Applications of Artificial Intelligence*, 2012.

[22] P. Mhaskar, N.H. El-Farra, and P.D. Christofides. Predictive control of switched nonlinear systems with scheduled mode transitions. *IEEE Transactions on Automatic Control*, 50(11):1670–1680, 2005.

[23] H. Michalska and D.Q. Mayne. Robust receding horizon control of constrained nonlinear systems. *IEEE Transactions on Automatic Control*, 38(11):1623–1633, 1993.

[24] S.J. Qin and T.A. Badgwell. A survey of industrial model predictive control technology. *Control engineering practice*, 11(7):733–764, 2003.

[25] R. Scattolini. Architectures for distributed and hierarchical model predictive control-a review. *Journal of Process Control*, 19(5):723–731, 2009.

[26] Brett T Stewart, Stephen J Wright, and James B Rawlings. Cooperative distributed model predictive control for nonlinear systems. *Journal of Process Control*, 21(5):698–704, 2011.

[27] M. Vaccarini, S. Longhi, and M.R. Katebi. Unconstrained networked decentralized model predictive control. *Journal of Process Control*, 19(2):328–339, 2009.

[28] A. Venkat, J. Rawlings, and S. Wright. *Distributed model predictive control of large-scale systems*. Springer, 2007.

[29] Aleksandar Zecevic and Dragoslav D Siljak. *Control of complex systems: Structural constraints and uncertainty*. Springer, 2010.

[30] Y. Zheng, N. Li, and S. Li. Hot-rolled strip laminar cooling process plant-wide temperature monitoring and control. *Control Engineering Practice*, 21(1):23–30, 2013.

[31] Y Zheng and S. Li. The stabilization of coordinated distributed model predictive control. In *Large scale complex systems theory and applications*, volume 13, pages 182–187, 2013.

[32] Y. Zheng, S. Li, and N. Li. Distributed model predictive control over network information exchange for large-scale systems. *Control Engineering Practice*, 19:757–769, 2011.

[33] Y. Zheng, S. Li, and H. Qiu. Networked coordination-based distributed model predictive control for large-scale system. *IEEE Transactions on Control Systems Technology*, 21(3):991–998, 2013.

[34] Y. Zheng, S. Li, and X. Wang. Distributed model predictive control for plant-wide hot-rolled strip laminar cooling process. *Journal of Process Control*, 19(9):1427–1437, 2009.

[35] Y. Zheng, S. Li, and X. Wang. Horizon-varying model predictive control for accelerated and controlled cooling process. *Industrial Electronics, IEEE Transactions on*, 58(1):329–336, 2011.

[36] Y. Zheng, S. Li, and X. Wang. Optimization target resetting distributed model predictive control for accelerated cooling process. In *Intelligent Control and Automation (WCICA), 2012 10th World Congress on*, pages 2764–2769. IEEE, 2012.

In: Model Predictive Control
Editor: Corrine Wade, pp. 145-154

ISBN: 978-1-63463-859-3
© 2015 Nova Science Publishers, Inc.

Chapter 7

FEASIBLE SET IN MODEL PREDICTIVE CONTROL

A. I. Propoi*
Institute of Systems Analysis,
Russian Academy of Sciences, Moscow, Russia

Abstract

A set of feasible trajectories for a control system is analyzed. We consider real trajectories (realized by a given control system or calculated by a computer) and possible trajectories (subjected to the equations and constraints of a given mathematical model). Therefore we have in model predictive control (MPC) the space of real trajectories and the space of possible trajectories. It is shown that the basic notion for the space of possible trajectories is an equivalence relation and associated with this relation a set of feasible trajectories (a ball in a Banach space). This ball with some additional requirements of symmetry determines a positive inner product; thus the space of possible trajectories becomes a Hilbert space (an Euclidean space for finite dimensional space). It is shown that the basic notion for the space of real trajectories is an order relation and associated with this relation a positive cone. Under some additional requirements of symmetry the cone becomes an elliptical cone that determines a pseudo-Euclidean metric in the space of real trajectories. In the last part of the paper the space of trajectories is considered as a direct sum of two spaces introduced above. A control possibilities set is introduced in this space. It is shown that the projections of this set onto the above spaces define the structure of control in MPC. This structure is described by the composition of two multivalued mappings: a given trajectory to a set of possible trajectories and a set of possible trajectories to a new (optimal) trajectory called pilot transformation. We describe pilot transformation by using the basic constructions of frame theory.

Keywords: model predictive control, feasible set, order and equivalence relations, production possibilities set, frame theory

AMS Subject Classification: 93-02, 92C10, 93D15, 42C15, 42A38

*E-mail address: propoi@gmail.com

1. Introduction

In model predictive control (MPC) the basic notions are a trajectory and a set of feasible trajectories.

We distinguish a real trajectory that has been realized by control system (CS) and a possible trajectory that can be realized by CS. The realization of a trajectory means the choice and the implementation of a sequence of control actions. We will also call a real trajectory the trajectory that has been computed from a mathematical model of CS for a given control.

On the other hand, possible trajectories are not fixed: they are only limited by the possibilities of a given CS.

We introduce the space of real trajectories and the space of possible trajectories. It is shown that the principal feature of the space of real trajectories is an order relation and the principal feature of the space of possible trajectories is an equivalence relation. Therefore, the structure of the feasible set is different in these two spaces. In the space of possible trajectories the feasible set is a convex symmetric body (an "ellipsoid") that defines Euiclidean metric for the estimation of the equivalence relation. In the space of real trajectories the feasible set is a positive cone that defines an order relation and pseudo-Euiclidean metric for the estimation of the order relation.

We show that these two types of the feasible set can be united in the notion of control possibilities set defined in the product of the possible and real trajectories spaces. (The definition of control possibilities set is similar to production possibilities set in mathematical economics).

Finally we show that the MPC procedure can be described as a composition of two mappings: a given current state (a real trajectory) to a feasible set (of possible trajectories) and a feasible set of possible trajectories to a new real trajectory. We call the composition of these two multivalued mappings pilot transformation [1]. In this chapter we describe pilot transformation by using the basic constructions of frame theory.

2. Control System

We give the following definition of control system (CS).

Definition 1. *Control system at each current state:*
generates a set of feasible states,
chooses a new state from this set.

We call a set of trajectories generated by CS at a time t the feasible set S_t.

In this definition and further we use "state" and "trajectory" as synonyms.

Two consequences follows from this definition.

First, feasible states are predictive by their nature: they can be realized by CS after the given state; second, to choose a new state the states from the feasible set should be evaluated.

From the definition it follows that controllable motion can be described as a composition of two multivalued mappings: state to feasible set and feasible set to state. This composition was introduced in [1] and called a pilot transformation.

In model predictive control (MPC) basic notions are a trajectory and a set of feasible (predictive) trajectories.

We will distinguish a real trajectory that has been (or is being) realized by CS and a possible trajectory that can be realized by CS.

Therefore there are two types of the trajectory space: the space of real trajectories and the space of possible trajectories.

We show that these spaces are different in their structure: the basic notion of the space of possible trajectories is the relation of equivalence and associated with this relation the Euclidean metric; the basic notion of the space of real trajectories is the relation of order and associated with this relation the pseudo-Euclidean metric.

3. The Space of Real Trajectories

We begin with the consideration of motion in a smooth n-dimensional manifold X. Let x be a point of this manifold. We consider the point x as a state of CS and X as the state space.

CS can be described by the differential inclusion:

$$\dot{x} = V(x), \tag{1}$$

where $V(x)$ is the set of possible velocities v at the point x, $v \in T_xX$, T_xX is the tangent space of the manifold X at the point x.

In particular, let

$$\dot{x} = f(x, u), \; u \in U, \tag{2}$$

then

$$V(x) = \{v \mid v = f(x, u), \; u \in U\}. \tag{3}$$

Let $\gamma(x_0, T)$ be a curve of the manifold X that CS has realized from the point x_0 by the choice of control actions $u(t)$, $t \in [0, T]$.

We call this curve a real trajectory. Here "real" means "realizable" and realization means the specification of control actions. Thus, a real trajectory either has been realized by CS, or has been specified by the choice of control.

Let $R_T(x_0)$ be the feasible set for CS (1) – (3). It consists of all points that can be achieved by CS for the time period $[0, T]$.

We omit the time parametrization of the feasible set and define in the manifold X the set of feasible trajectories $C(x)$ from the point x. Now we can define a real trajectory in a more formal way (see also [2]).

Definition 2. *We call a real trajectory a curve γ of the manifold X that satisfies the condition: if $x \in \gamma$, then exists such a point $x' \in \gamma$ that $x' \in C(x)$.*

Note that the curve γ in definition 2 is not parameterized.

From definition 2 follows that the points of real trajectories are obeyed an order relation. We remind the definition [3].

Definition 3. *A binary relation R is called order if*

$$aRa\ (reflexivity), \tag{4}$$

$$aRb\ and\ bRc \Rightarrow aRc\ (transitivity), \tag{5}$$

$$aRb\ and\ bRa \Rightarrow a = b\ (symmetry). \tag{6}$$

If the relation R satisfies only (4) and (5), the R is the relation of preorder (or quasi-order).

Proposition 1. *The set of feasible trajectories has properties:*

$$x \in C(x)\ (reflexivity), \tag{7}$$

$$x_1 \in C(x_0)\ and\ x_2 \in C(x_1) \Rightarrow\ x_2 \in C(x_0)\ (transitivity). \tag{8}$$

We prove only transitivity. Let γ_x be the feasible trajectories from x.

If $x_1 \in C(x_0)$ then a feasible trajectory γ_0 exists such that $x_1 \in \gamma_0$. If $x_2 \in C(x_1)$ then a feasible trajectory γ_1 exists such that $x_2 \in \gamma_1$. The union of these two trajectories gives the feasible trajectory from x_0 to x_2 because a point $x_0' \in \gamma_0$ exists such that $x_1 \in C(x_0')$ and a point $x_1' \in \gamma_1$ exists such that $x_1' \in C(x_1)$.

From definition 3 and proposition 1 follows that there exists a relation of preorder on the set of real trajectories X. We denote this relation as $\geq$. With the relation of preorder $\geq$ is associated a positive cone

$$K(x) = \{x \,|\, x \geq 0\}. \tag{9}$$

An analysis of the above constructions shows that in fact they hold not only for the points of (real) trajectories but for the trajectories themselves.

From now on we will mean that x is a trajectory and call x the state of CS; $C(x)$ is the set of feasible trajectories at the state x.

We introduce the relation:

Relation 1. *The state x' is feasible for the state x*

and will understand here "feasible" as "realized". In other words, CS moving along a trajectory x realizes (at some time t) a new trajectory x'.

If $C(x)$ is the set of all feasible trajectories at the state x, then this relation can be written as $x' \in C(x)$. It was shown that this relation is a preorder and, consequently, $x' \in C(x)$ means $x' \geq x$ (i.e., $C(x)$ is a positive cone at x).

The nature of controlable motion justifies the assumption:

Assumption 1. *If $x \geq x'$ and $x' \geq x$, then $x' = x$.*

In this case $\geq$ is a relation of order.

Thus, the set of real trajectories is a partially ordered set (poset): all trajectories $x' \in C(x)$ (i.e., $x' \geq x$) can be realized by CS at (after) the state x.

For applications it is important to describe this relation in geometric language. It will be done in section 5.

4. The Space of Possible Trajectories

If the basic feature of the space of real trajectories is motion (i.e., a change of states), then the basic feature of the space of possible trajectories is the process of generation (of trajectories).

By the nature of controllable motion all trajectories generated by CS at a given state (that is, before the choice of an optimal trajectory) are equivalent: they are only limited by characteristics of a given CS (and, thus, define a feasible set) but they are not specified at this stage). We again introduce

Relation 2. *The state x' is feasible for the state x*

but now we will understand "feasible" as "possible". In other words, CS at a given (real) state x can generate a trajectory x'.

We denote the set of possible trajectories as $B(x)$.

The process of generation justifies the following assumption.

Assumption 2. *The set $B(x)$ satisfies the conditions:*

$$x \in B(x)\ (reflexivity), \tag{10}$$

$$z \in B(x), y \in B(z)\ \Rightarrow\ y \in B(x)\ (transitivity), \tag{11}$$

$$y \in B(x)\ \Leftrightarrow\ x \in B(y)\ (symmetry). \tag{12}$$

From this assumption follows that the relation 2 is a relation of equivalence [3]. We denote it as $x' \sim x$.

The feasible set $B(x)$ is an equivalence class with the representative x. This set contains all trajectories x' equivalent to x:

$$B(x) = \{x' \,|\, x' \sim x\}. \tag{13}$$

Thus, the basic notion of the set of possible trajectories is the relation of equivalence: all trajectories generated by CS at a given state are equivalent before the decision making (on the choice of an optimal trajectory).

In the next section we give geometric interpretation of these relations.

5. Geometry of Feasibility Set

In this section we consider the space of trajectories as a real Hilbert space X.

Let $x \in X$ be a trajectory of duration $[0, T]$. We define the feasible set of trajectories at the given state:

$$B(x) = \{u \,|(u, Gu) \leq 1\}, \tag{14}$$

where $u = x' - x$, G is a bounded positive self-adjoint linear operator $X \to X$, (u, v) is the inner product in X.

The quadratic form in (14) induces a new metric in X: $(x, Gx) = \|x\|_G^2$; $B(x)$ is the unit ball in this metric. Note that in general case the ball $B(x)$ lies in the feasible subspace S_x of the space X (it is a disk in X).

Proposition 2. *The set $B(x)$ in (14) satisfies the properties:*

$$x \in B(x) \ (reflexivity), \tag{15}$$

$$z \in B(x), y \in B(z) \ \Rightarrow \ y \in B_2(x) \ (transitivity), \tag{16}$$

$$y \in B(x) \ \Leftrightarrow \ x \in B(y) \ (symmetry). \tag{17}$$

The property (16) is a formalization of the Huygens principle of generation: the ball $B_2(x)$ contains trajectories of duration $[0, 2T]$.

From (17) follows the symmetry of the set $B(x)$:

$$u \in B(x) \ \Leftrightarrow \ -x \in B(x). \tag{18}$$

The properties (17), (18) reflect the principal (and nontrivial) feature of the states generations — its symmetry (opposite to the motion). Note that in (17), (18) the symmetry of trajectory generation is expressed in a stronger way than in (12): if a trajectory y can be generated at state x for the time period $[0, T]$, then a trajectory x can be generated at state y for the same period.

We now introduce the dual space X^* (the space of linear functionals on X). Let $p \in X^*$ evaluate the trajectory u. We will consider the value (p, u) as a criterion for the choice of a trajectory from the feasible set. Then this choice is reduced to the maximization problem

$$\max\{(p, u) \,|\, u \in B(x)\} = \max\{(p, u) \,|\, (u, Gx) \leq 1\}. \tag{19}$$

Let

$$H_\beta = \{u \,|\, (p, u) = \beta\} \tag{20}$$

be a hyperplane in X.

If $\beta = 1$, then (20) is the support hyperplane for the ball $B(x)$ at u_p. If $0 < \beta < 1$, then this hyperplane defines a cross-section of the ball $B(x)$: $F_\beta = B(x) \cap H_\beta$ (it is a ball (disk) in H_β).

The frontier of the section is defined by relations

$$\bar{F}_\beta = \{u \,|\, \|u\|_B = 1, \ (p, u) = \beta\}. \tag{21}$$

The equalities in (21) define two equivalent classes of trajectories: the first one with respect to the generation of trajectories and the second one with respect to the goal of control.

As a result, we obtain a cone of feasible trajectories for the state x:

$$K_\beta(x) = \{tu \,|\, u \in F_\beta, \ t \geq 0\}. \tag{22}$$

Let $x = tv$, $W = \text{Ker}\ (p) = \{w \,|\, (p, w) = 0\}$. Then the space X can be decomposed into the orthogonal sum $X = X_v \oplus W$. In this decomposition the cone can be defined as

$$K(v) = \{(t; w) \,|\, \|w\|^2 \leq t^2 v^2, \ t \geq 0, \ w \in W\}. \tag{23}$$

The cone (23) is an elliptic cone. It defines the pseudo-Euiclidean metric (the Minkowskii metric) for the estimation of an order relations [4],[5].

If in (23) t is fixed, then (23) defines the ball in W of radius tv — the section of the ball (14) in X by the hyperplane (20).

The cone (22), (23) is a positive cone. Therefore, it defines a relation of order. Similarly, the ball $B(x)$ in X (or a ball in W) defines a relation of equivalence. Thus we linked the relations of order and equivalence introduced in sections 2 and 3 with the geometry of the feasible set. Note that these relations are consistent with vector and topology structure of the trajectory space.

6. The Set of Control Possibilities

We change slightly the notations. Let X be the space of real trajectories and Y be the space of possible trajectories. We define the state space of CS Z as the direct sum of these spaces; $Z = X \oplus Y$. We assume that the spaces X and Y have the same dimension and a common origin $x_0 = y_0 = 0$.

By definition, CS changes the current state x_0 to the (feasible) state x and generates at this state x the set of possible states y, then CS chooses and realizes a new state x' from these states. We formalize this process.

Definition 4. *We call the set*

$$T(x_0, y_0) = \{(x, y) \in Z \mid y \in B(x, x_0),\ x \in K(x_0)\} \tag{24}$$

the set of control possibilities of CS, or shortly control possibilities set.

Note that this definition is similar to the definition of the production possibilities set in mathematical economics [6], [7].

In the language of binary relations

$$T(x_0, y_0) = \{(x, y) \mid y \sim x,\ x \geqslant x_0\}. \tag{25}$$

Verbally: the state x is feasible (realizable) at the state x_0, and the state y is feasible (possible) at the state x.

We fix x_0 in (24) and assume that $x_0 = y_0 = 0$. The corresponding set will be denoted as T.

In this case

$$T = \{(x, y) \in Z \mid y \text{ is feasible from } x\}, \tag{26}$$

or, the input x can produce the output y (cf relations 1 and 2).

We assume that T is a positive cone in Z; besides, we assume that if $(x, y) \in T$, then $(x', y') \in T$ for $x' \geq x$, and $y \sim x'$.

The sets (24) or (26) induce two multivalued mappings: $X \to Y$ and $Y \to X$, each of these mappings is the inverse of the other.

These mappings are defined by the cross-sections of the set T:

$$B(x) = \{y \mid (x, y) \in T\}, \tag{27}$$

$$D(y) = \{x \mid (x, y) \in T\}. \tag{28}$$

Let $z = w + tv + \tau e = x + \tau e = y + tv$, where $x = w + tv$, $y = w + \tau e$, or $z = (w, t, \tau) = (x, \tau) = (y, t)$. In this presentation the cross-sections (27), (28) of the cone $T(z_0)$, $z_0 = (x_0, y_0)$, can be described as

$$B_t(x_0) = \{(w, \tau) \mid (w, t, \tau) \in T(z_0)\} = \{y \mid (y, t) \in T(z_0)\}, \tag{29}$$

$$D_\tau(y_0) = \{(w, t) \mid (w, t, \tau) \in T(z_0)\} = \{x \mid (x, \tau) \in T(z_0)\}. \tag{30}$$

Under symmetry conditions the set (29) is an Euclidean ball and the set (30) is a pseudo-Euclidean ball (it is a cone for $t = 0$ [4],[5]).

Thus, the control possibilities set unites the properties of the feasibility sets in the real trajectories space and in the possible trajectories space.

7. Pilot Transformation

As was stated in section 2, controlling motion includes two stages: the generation of possible trajectories at a given state, and the choice of a new (optimal) trajectory from this set.

Formally, it is a composition of two mappings: a given current state (a real trajectory) to a feasible set (of possible trajectories) and a feasible set of possible trajectories to a new real trajectory, We call it pilot transformation [1]. In this section we describe pilot transformation by using the basic constructions of frame theory [8].

Let $\{x_j\}$, $j \in J$, be the set of countable trajectories generated by CS at the state x_0. Each of these trajectories is determined by a concrete input (control), so $x(j)$ is a real (feasible) trajectory in the sense of definition of section 3. It is assumed that $x(j) \in H$, H is a Hilbert space.

We assume that this set of trajectories satisfies the following inequalities

$$A\|x\|^2 \leq \sum_{j \in J} (x(j), x)^2 \leq B\|x\|^2. \tag{31}$$

Such a collection of vectors is called frame X, where X is the matrix of vectors $\{x_j\}, j \in J$ (and the corresponding operator $X : H \to H$) [8].

Note that the first inequality in (31) means that the trajectories $\{x_j\}$, $j \in J$, span the feasible subspace of CS S in H.

It is supposed in the frame theory that the evaluation of a trajectory x with respect to a frame $\{\tilde{x}_j\}$, $j \in J$, is performed by using the inner product

$$(\tilde{x}(j), x) = z(j), \tag{32}$$

where the trajectories $\tilde{x}(j)$ and $x(j)$ are connected by some invertible transformation (e.g., they are dual).

In the matrix form (32) becomes

$$\tilde{X}^* x = z. \tag{33}$$

So the first stage of controlling motion can be described by the following chain: $x_0 \to \{x(j)\} \to \{\tilde{x}(j)\} \to \{z(j)\}$, or as a mapping $H \to l^2(J)$, where $l^2(j)$ is the space of square summable numbers $\{z_j\}, j \in J$. In the frame theory this stage is called measurement or sampling.

Thus, CS at a given state x_0 generates a set of feasible trajectories $\{x(j)\}$.

The second stage, e.i., the choice of a new (optimal) trajectory can be described by the mapping $l^2(J) \to H$:

$$x = \sum_{j \in J} x(j) z(j), \tag{34}$$

or, in matrix form

$$x = Xz. \tag{35}$$

We call this stage the construction stage (of a new trajectory).

The composition of these two mappings $H \to l^2$ and $l^2 \to H$ defines the pilot transformation

$$x = \sum_{j \in J} x(j)\tilde{x}^*(j)x_0 = X\tilde{X}^* x_0. \tag{36}$$

The order of using frames X and $\tilde{X}$ can be changed

$$x = \tilde{X}X^* x_0. \tag{37}$$

If frames X and $\tilde{X}$ are dual:

$$\tilde{x}(j) = (X\tilde{X}^*)^{-1} x(j), \tag{38}$$

then (36) becomes

$$x = \sum_{j \in J} x(j)\tilde{x}^*(j)x = X\tilde{X}^* x. \tag{39}$$

In this case CS reconstruct the current trajectory $x = x_0$, i.e. continues its motion along x_0. (For more details of application of frame theory in MPC see [9].)

8. Conclusion

In this chapter we have briefly analyzed the basic features of the feasible set in MPC. It was shown that the feasible set can be considered either as an ellipsoid of possible trajectories or as a positive cone of feasible (realizable) trajectories. These two types of the feasible set are united in the notion of control possibilities set. The analysis reveals deep relations of the feasible set to fundamental notions in other sciences: the control possibilities set is similar to the production possibilities set in mathematical economy which is a basis for the study of growth and equilibrium properties of a system [6], [7]; the positive cone of feasible trajectories is similar to elliptic cone in the relativity theory which is a basis for the study of the cone transformations [10].

References

[1] Propoi, A.I, (2007), On the structure of controlling motion. *Nonlinear Dynamics and Control*, pages 59 –68, M.: Fizmatlit.

[2] Auben, J.-P., and Ekeland, K. (1984), *Applied Nonlinear Analysis*, NY: Wiley-Interscience.

[3] Birkhoff, G., and Mac Lane, S. (1999), Algebra, 3rd edition, Chelsea.

[4] Propoi, A.I., (2004), On constructing a metric on a dominating cone. I, *Automation and Remote Control*, 3, 49 - 58.

[5] Propoi, A.I., (2004), On constructing a metric on a dominating cone. II, *Automation and Remote Control*, 4, 52 - 60.

[6] Carter, M., (2001), *Foundations of Mathematical Economics*, BO: MIT Press.

[7] Bewley, T., F. (2007), *General Equilibrium, Overlapping Generations Models, and Optimal Growth Theory*, Harvard University Press.

[8] Christensen, O. (2003), *An Introduction to Frames and Riesz Bases*. Boston, MA, U.S.A.: Birkhauser.

[9] Propoi, A. I. (2015), Model predictive control in Hilbert space, *Chapter in this volume*.

[10] Naber, G. (2012), *The geometry of Minkowski Spacetime*, B: Springer.

INDEX

N

O

P

Q

R

S

T

U

V

W